43 iwe 428
lbf 538 03.Exp.

Ausgeschieden im Jahr 2025

Signals and Communication Technology

For further volumes:
http://www.springer.com/series/4748

Stephan Pachnicke

Fiber-Optic Transmission Networks

Efficient Design and Dynamic Operation

Springer

Dr.-Ing. Stephan Pachnicke
Dipl.-Wirt.-Ing.
Chair for High Frequency Technology
TU Dortmund
Friedrich-Wöhler-Weg 4
44221 Dortmund
Germany
e-mail: stephan.pachnicke@tu-dortmund.de

ISSN 1860-4862
ISBN 978-3-642-21054-9   e-ISBN 978-3-642-21055-6
DOI 10.1007/978-3-642-21055-6
Springer Heidelberg Dordrecht London New York

Library of Congress Control Number: 2011936529

© Springer-Verlag Berlin Heidelberg 2012
This work is subject to copyright. All rights are reserved, whether the whole or part of the material is concerned, specifically the rights of translation, reprinting, reuse of illustrations, recitation, broadcasting, reproduction on microfilm or in any other way, and storage in data banks. Duplication of this publication or parts thereof is permitted only under the provisions of the German Copyright Law of September 9, 1965, in its current version, and permission for use must always be obtained from Springer. Violations are liable to prosecution under the German Copyright Law.
The use of general descriptive names, registered names, trademarks, etc. in this publication does not imply, even in the absence of a specific statement, that such names are exempt from the relevant protective laws and regulations and therefore free for general use.

Cover design: eStudio Calamar, Berlin/Figueres

Printed on acid-free paper

Springer is part of Springer Science+Business Media (www.springer.com)

Preface

Next generation optical communication systems will be characterized by increasing data rates, dynamic adaptation to the actual traffic demands and improved energy efficiency. In future all-optical meshed networks it is highly desirable that transparent optical paths can be set up on-the-fly, providing bandwidth-on-demand and shutting down unused resources automatically.

Appropriate design of an optical core network is crucial as the capital expenditure for such a system is extremely high. Currently the split-step Fourier method is employed in state-of-the-art transmission system simulators to solve the nonlinear Schrödinger equation describing the propagation of light in a single mode glass fiber. Unfortunately, even on today's computer systems the simulation of a wavelength division multiplex transmission system with a high channel count and a transmission distance of several thousand kilometers is still very time-consuming with computational efforts of hours to several days. This is why in this book several alternative methods are proposed to reduce the simulation time. They range from a meta-model based optimization method for finding an optimum operation point of a transmission system in a large parameter space with only a very low number of numerical simulations to parallelization techniques for solving the nonlinear Schrödinger equation in an efficient way on a graphics processing unit (GPU) with a high number of cores. Alternatively, fast analytical models for the approximation of various degradation effects occurring on the transmission line can be used.

The design of fiber optical core networks is evolving rapidly from static point-to-point transmission systems to meshed networks with transparent optical cross-connects enabling dynamic reconfiguration of the network and leading to longer transparent transmission distances. During recent years several research projects have been conducted throughout the world investigating on-the-fly provisioning of bandwidth with demand granularities in the order of a wavelength capacity. To exploit the maximum transparent reach the inclusion of physical-layer impairments in the routing and wavelength assignment process is desirable. In this book constraint-based routing and regenerator placement are presented based on fast analytical models for approximating the signal quality. It is shown that the blocking probability of dynamic wavelength demands as well as the number of required optical-electrical-optical regenerator components can be reduced significantly by the inclusion of physical-layer impairments. Finally, it is investigated how the energy efficiency of an optical core network can be increased considerably by making use of optical bypasses to keep more traffic entirely in the optical layer and shutting down unused components in the IP-layer.

Most of the work presented here has been carried out in close collaboration with industry partners during my time at the Chair for High Frequency Technology of TU Dortmund, Germany. The parallelization techniques on GPUs have already been included in the commercially distributed optical system simulator *PHOTOSS*. Network operation research has lead to the development of a new simulation tool named *CBRSuite*.

Acknowledgments

First and foremost, I wish to acknowledge sincerely Prof. Peter M. Krummrich for his help throughout the whole project and for his support and encouragement to complete this work.

I would also like to acknowledge Prof. Edgar Voges, emeritus of the Chair for High Frequency Technology, who took part in this work passionately. Unfortunately, Prof. Voges passed away much too early in the end of the year 2008 so he could not see the completion of this work anymore. He gave me valuable advice until his very end. I will always remember him for all his contributions and as a role-model of a caring professor.

For their extensive support in implementing the simulation tools *PHOTOSS* and *CBRSuite*, I would like to thank my collaborators Nicolas Luck, Christian Remmersmann, Daniel Teufer and Matthias Westhäuser as well as the involved students and software developers. It should be mentioned at this point that many of the GPU routines presented in this book have been implemented by Adam Chachaj. Many thanks go also to my former colleagues, especially to Martin Windmann for his permanent help in day-to-day and special matters.

Special thanks go to Petra Sauerland for many technical drawings and to Iris Wolff for taking over administrative tasks.

Dortmund, November 2011   Stephan Pachnicke

Contents

1 **Introduction** 1
   1.1 Historical Evolution 1
   1.2 Motivation 4
   1.3 Structure of This Book 5
   References 6

Part I Optical Network Design

2 **Fiber Optical Transmission Systems** 11
   2.1 Generic Setup 11
   2.2 Transmitters 13
   2.3 Modulation Formats 15
   2.4 Fiber Properties 18
   2.5 Amplifiers 21
   2.6 Optical Cross Connects 22
   2.7 Receivers 24
   2.8 Electrical Signal Processing 26
   References 27

3 **Simulation of Fiber Optical Transmission Systems** 31
   3.1 Modeling of Fiber Optical Transmission Systems 31
      3.1.1 Attenuation and Dispersion 32
      3.1.2 Amplified Spontaneous Emission Noise 35
      3.1.3 Fiber Birefringence 38
      3.1.4 Nonlinear Fiber Effects 39
      3.1.5 Coupled Nonlinear Schrödinger Equation 42
      3.1.6 Split-Step Fourier Method 44
      3.1.7 Modeling of Polarization Mode Dispersion 46
      3.1.8 Calculation of the Bit Error Ratio 48

		3.2	The Fiber Optical Transmission Simulator PHOTOSS	50
	References			52
4	**Efficient Design of Fiber Optical Transmission Systems**			55
	4.1	Meta-Heuristic Based Optimization		56
		4.1.1	Overview of Employed Algorithms	57
		4.1.2	Meta-Model	61
		4.1.3	Analysis of Exemplary Transmission Systems	63
	4.2	Parallelization of a Simulation on a Graphics Processing Unit		67
		4.2.1	Implementation of the FFT and Split-Step Fourier Method on a GPU	68
		4.2.2	Stratified Monte-Carlo Sampling Technique	78
	4.3	Analytical Modeling of the Signal Quality		83
		4.3.1	Linear Degradation Effects	84
		4.3.2	Nonlinear-Degradation Effects	87
		4.3.3	System Example	94
	4.4	Summary and Discussion		97
	References			99

Part II Optical Network Operation

5	**Dynamic Operation of Fiber Optical Transmission Networks**			105
	5.1	Network Architecture		106
	5.2	Demand Model		109
	5.3	Constraint-Based Routing and Regenerator Placement		111
		5.3.1	Assessment of the Signal Quality by a Single Figure of Merit	111
		5.3.2	Physical Layer Impairment Aware Routing Algorithm	114
		5.3.3	Regenerator Placement Heuristic	116
		5.3.4	Results	118
		5.3.5	Reduction of the Required Number of Electrical Regenerators	121
	5.4	Extensions to High Bit Rate Systems with Novel Modulation Formats		125
	5.5	Improvement of Energy Efficiency		129
		5.5.1	Power Consumption of Deployed Components	131
		5.5.2	Grooming	132
		5.5.3	Approach for Reducing Core Network Energy Consumption	133
		5.5.4	Examplary Study	134
		5.5.5	Reduction of Energy Consumption by Load-Adaptive Operation	138

	5.6	Summary and Discussion	140
		References	142
6	**Conclusions and Outlook**		147
		References	151

Appendix ... 153

Index ... 159

Chapter 1
Introduction

Abstract After a brief discussion of the expected network traffic growth and the historical evolution of optical network capacity over the last 25 years, the need for agile and energy efficient optical networks is pointed out. Furthermore, it is motivated why efficient (numerical) simulation and optimization of fiber-optic transmission systems are highly desirable and which approaches are promising to achieve these goals. Finally, the scope of this book is outlined and the contents of the different chapters are summarized.

1.1 Historical Evolution

Telecommunication has become a central part of our life within the last century. Especially in the last 20 years disruptive changes have taken place with the emergence of the Internet [1] and ubiquitous availability of mobile telecommunications in many areas around the globe. The rapid development of the telecommunication sector can also be seen by an exponential growth of the data traffic being transported [2]. Currently, the annual growth rate of the data volume is in the range of 50–100% [3]. As network traffic is dominated by Internet traffic, a good indicator for the transported traffic volume is the average traffic in a large Internet exchange such as DE-CIX in Frankfurt, Germany (compare Fig. 1.1).

The boost of communications requires increasing data rates and broadband access to the Internet. Through the introduction of digital subscriber line (DSL)[1] households are today able to download broadband contents such as high-definition movies in real-time. The Internet services have also transformed from mostly static text-oriented to interactive multimedia services. Websites such as *youtube.com,* or

---

[1] A complete list of abbreviations used in this work can be found in the appendix.

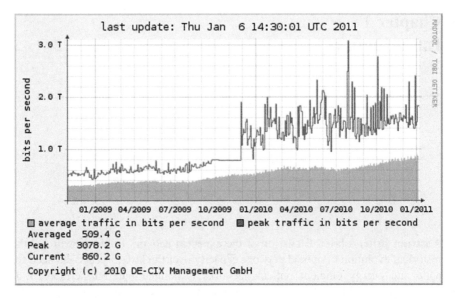

Fig. 1.1 Network traffic growth at DE-CIX Internet exchange in Frankfurt [3]

flickr.com as well as social media networks such as *facebook.com* are widely known today. Furthermore, cloud computing is maturing and e.g. web space services such as Amazon's simple storage service (S3) already require high data volumes to be transported.

Also legacy services such as the plain old telephone system are evolving, and in a few years voice traffic will most likely be transported exclusively using the packet-oriented voice-over-IP (VOIP) protocol.

Currently there are emerging trends in Germany to further increase the data rates of the subscribers and to roll out a high speed fiber optical access network to the end user providing even higher data rates of more than 100 Mb/s to a few Gb/s [4, 5].

For cost-effective transport of large amounts of data over long-distances almost exclusively optical networks using fiber optical waveguides are utilized. In today's backbone networks solely single-mode fibers are deployed, which are produced from silica glass. The attenuation of a single-mode silica fiber has a minimum around the wavelength of 1,550 nm ($\alpha \approx 0.2$ dB/km), and its available bandwidth is extremely high. In the low-loss window (from 1,460 to 1,625 nm) a total bandwidth of more than 20 THz can be utilized.

In the past 25 years optical (long-haul) communications have moved from single-channel systems carrying about 1 Gb/s to today's commercially available transmission systems with 100 Gb/s per channel (wavelength) [6]. Present-day commercial systems offer wavelength counts of up to 160 wavelength division multiplex (WDM) channels. A crucial innovation was the Erbium-doped fiber amplifier (EDFA) in the late 1980's [7]. It enables to bridge long-distances without electro-

1.1 Historical Evolution

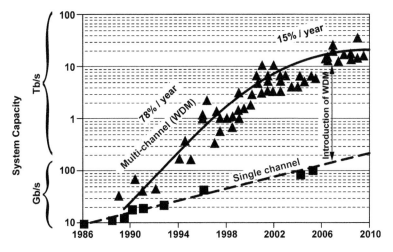

Fig. 1.2 Historical evolution of optical network capacity (adapted from [10])

optical conversion. Furthermore, a high number of WDM channels can be amplified simultaneously using only one EDFA. During the 1990's the capacity-distance product [8] as a measure for the technical accomplishment was continuously enhanced. In spite of the tremendous growth in transmitted capacity (compare Fig. 1.2), the large exponential growth of the traffic volume leads to the question of the ultimate capacity of optical communication fibers. For linear channels the solution is well known [9]. For nonlinear channels—such as optical fibers in long-haul communications—the capacity is reduced, however. Recent estimates of the ultimate capacity have shown that a spectral efficiency per polarization of about 9 bits/s/Hz is achievable over 500 km of standard single-mode fiber [10]. This means that roughly a one order of magnitude increase in spectral efficiency remains in the future for commercial optical communication systems.

In contrast to electrical systems, optical fiber communication systems are still at the very beginning of a commercially practical implementation of high-order modulation formats [11]. Through the adoption of higher-order modulation formats, higher spectral efficiencies will be reached by the reduced symbol rate. The key technology for advanced optical modulation formats is coherent detection, which however increases the complexity of the receivers significantly compared to the most widely deployed direct detection at the receiver end. Coherent detection is very beneficial because it makes available all the optical field parameters in the electrical domain. Only coherent detection will permit convergence to the ultimate limits of spectral efficiency [12].

Apart from the pure increase of transmitted capacity also commercial networking services have changed radically over the last 25 years. In today's networks the traffic is dominated by video and data with voice traffic representing only a small percentage of the total traffic volume. The opposite was true in the mid 1980's [13]. This also led to changes in the network agility. Traditionally (long-haul) networks have been fairly static with connections remaining established for months or even years. With the development of new services such as e.g. grid computing, however, high bitrate pipes are required for relatively short periods of time, and the network must be rapidly reconfigurable [14]. It is clear that the steady traffic growth with a greater diversity of services will continue in the future. To meet this growth, carriers will continue to seek technologies that provide cost, scalability, and operational advantages.

The intensive public discussion on climate change has been attracting more and more interest in telecommunication network energy efficiency in recent years [15]. The telecommunication network energy consumption is currently growing exponentially as a result of increasing numbers of users and data volume. As energy costs contribute significantly to the operational expenditures of telecommunication network operators, energy efficiency improvements have become a vital economic interest of many network operators today. In the future the power consumption of the core network is expected to exhibit the highest growth rates because the data volume is increasing more quickly than the number of users. This makes improvements in core network energy efficiency highly desirable.

1.2 Motivation

Despite the high transmission bandwidth increase in the last decades network operators are still under a high pressure to lower the cost per transported bit/s/km. To reach this goal both the system design and also network operation must be optimized.

During the design phase numerical simulations play a major role. They are used to identify the optimum working point of a system. In simulations a large parameter space can be searched before the deployment of a new or upgraded system. Numerical simulations of optical fiber links are typically based on the split-step Fourier method (SSFM), which unfortunately requires a considerable computational effort for the typically required accuracy. Especially for transmission systems with a high number of WDM channels, with significant nonlinear effects in the fiber and with a large transmission distance the computational time for propagation of the signal through the link takes several hours or even days on a state-of-the-art desktop computer, which is the standard equipment for a system designer.

To improve the design phase a more efficient numerical simulation scheme is thus desirable. For this purpose several different approaches can be chosen. One possibility is to reduce the number of required numerical simulations (or samples) for identifying the optimum working point. Due to the complexity of the problem especially heuristic approaches are advisable. Another possibility to reduce the computational time of the numerical simulations is to parallelize them e.g. on a computer with a high number of processing cores. A special form of a multi core processor is a graphics card, which also facilitates a high number of cores. An additional benefit of graphics cards is that the price per core is currently much lower compared to CPUs making GPUs preferable in many cases.

1.2 Motivation

Apart from the numerical simulations with the SSFM it is also possible to utilize analytical or semi-analytical models to approximate the signal quality along a transmission link. These models have the advantage of being very fast, however, at the expense of generic applicability. Most models have to be adapted to a specific modulation format and bit rate.

To reduce costs and to facilitate new applications and services also network operation has to be enhanced continuously. During the operation of optical core networks a dynamic and configurable optical layer, which is able to serve dynamic wavelength requests, is envisoned for the future [13]. Lightpaths (wavelengths with a certain bandwidth capacity) may be requested by the user on-demand. Photonic paths are not currently common service offerings, but they will offer significant infrastructure benefits such as reduced cost, space, and power dissipation. To setup such dynamic wavelength demands new physical-layer impairment (PLI) aware routing algorithms are needed, which allow exploiting the maximum amount of transparency.

In the recent past the field of energy efficiency of network equipment has moved into the focus of many research efforts. Especially concerns have been raised that the energy consumption of information and communication technology equipment is lagging network traffic growth rates. It is of high interest which countermeasures provide the best opportunities to increase energy efficiency.

Both fields, optimization of optical network design and improved dynamic energy efficient operation, are discussed.

1.3 Structure of This Book

This book is organized as follows. In Chap. 2 the basic configuration of a fiber optical transmission system is outlined. Furthermore, the different components and their fundamental working principles are explained.

Chapter 3 deals with the simulation of fiber optical transmission systems. First, the most important effects occurring in silica glass fiber based transmission are presented. Mathematically the propagation of radiation in the glass fiber can be described by the nonlinear Schrödinger equation, which is discussed in detail. For solving this equation usually a split-step Fourier approach is used. The chapter is complemented by a short introduction into polarization effects occurring in a single-mode fiber and into the simulation methodology typically used to simulate polarization-mode dispersion (PMD).

The efficient design of fiber optical transmission systems is addressed in Chap. 4. Several approaches to reduce the simulation time are outlined. First, a novel meta-heuristic based optimization method is presented. This method can be used to find the optimum parameter set in a large parameter space without performing a time consuming grid search (i.e., varying certain system parameters in equidistant steps).

Furthermore, a new parallel implementation of the SSFM on a graphics processing unit (GPU) with a high number of processing cores is presented. It is shown how single- and double-precision-based simulations can be combined by a stratified Monte-Carlo sampling scheme. This method allows ensuring a predefined accuracy independent of the simulation parameters (i.e. the number of split-steps).

As the assessment of the signal quality in (almost) real-time is desirable for the operation of dynamic optical networks, analytical formulas are needed, which can approximate the signal quality on the fly. For this purpose a set of (semi-) analytical formulas and heuristics is presented, which allow to approximate the signal degradation caused by the various linear and nonlinear effects occurring along the transmission line.

Chapter 5 focuses on the dynamic operation of a fiber optical transmission network. In this chapter it is shown how the network performance can be improved by the use of physical-layer impairment-aware routing algorithms. Furthermore, a significant reduction in (costly) components such as regenerators can be achieved by including physical-layer performance in the design phase. As most analytical models are currently available only for transmission systems with a line rate of 10 Gb/s per wavelength and intensity modulation, possible extensions to novel modulation formats and higher line rates are suggested.

At the end of the chapter a study on the energy consumption of a fiber optical core network of pan-European dimensions is shown. In such a network it can be observed that the traffic load exhibits diurnal variations ranging from approximately 25% in the early morning to peak values in the evening. It is investigated how much energy can be saved by optical-bypassing and partial deactivation of network elements during low load periods.

Finally, in Chap. 6 conclusions are drawn and an outlook to some potentially interesting future research topics are given.

References

1. Abbate, J.: Inventing the internet (inside technology). MIT Press, Cambridge (2000)
2. Desurvire, E.: Optical communications in 2025. European Conference on Optical Communication (ECOC), Mo.2.1.3, Glasgow, Sept 2005
3. DE-CIX German Internet Exchange (www.decix.de)
4. Breuer, D., Monath, T.: Research questions in the business case of FTTH. European Conference on Optical Communication (ECOC), WS7, Torino, Sept 2010
5. Green, P.E.: Fiber to the home: the next big broadband thing. IEEE Comm. Mag. **42**(9), 100–106 (2004)
6. Gnauck, A.H., Tkach, R.W., Chraplyvy, A.R., Li, T.: High-capacity optical transmission systems. IEEE/OSA J. Lightw. Technol. **26**(9), 1032–1045 (2008)
7. Olsson, N.A.: Lightwave systems with optical amplifiers. IEEE/OSA J. Lightw. Technol. **7**(7), 1071–1082 (1989)
8. Henry, P.: Lightwave primer. IEEE J. Quant. Electron. **21**(12), 1862–1879 (1985)

References

9. Shannon, C. E.: A mathematical theory of communication. Bell. Syst. Tech. J. **27**, 379–423, 623–656 (1948)
10. Essiambre, R.-J., Kramer, G., Winzer, P.J., Foschini, G.J., Goebel, B.: Capacity limits of optical fiber networks. IEEE/OSA J. Lightw. Technol. **28**(4), 662–701 (2010)
11. Seimetz, M.: Higher-order modulation for optical fiber transmission. Springer, Berlin (2009)
12. Kahn, J.M., Ho, K.-O.: Spectral efficiency limits and modulation/detection techniques for DWDM systems. IEEE J. Sel. Top. Quant. Electron. **10**(2), 259–272 (2004)
13. Berthold, J., Saleh, A.A.M., Blair, L., Simmons, J.M.: Optical networking: past, present, and future. IEEE/OSA J. Lightw. Technol. **26**(9), 1104–1118 (2008)
14. Gladisch, A., Braun, R.-P., Breuer, D., Erhardt, A., Foisel, H.-M., Jäger, M., Leppla, R., Schneiders, M., Vorbeck, S., Weiershausen, W., Westphal, F.-J.: Evolution of terrestrial optical system and core network architecture. Proc. IEEE **94**(5), 869–891 (2006)
15. Lange, C., Kosiankowski, D., Weidmann, R., Gladisch, A.: Energy consumption of telecommunication networks and related improvement options. IEEE J. Sel. Top. Quant. Electron. **17**(2), 285–295 (2011)

Part I
Optical Network Design

Chapter 2
Fiber Optical Transmission Systems

Abstract In this chapter the basic concepts of fiber optical transmission systems are explained. The chapter starts with the presentation of the generic setup of a wavelength division multiplexing optical long-haul system. Afterwards the most important components are introduced, which are transmitters, optical amplifiers, fibers, optical cross-connects and receivers. At this point only the general properties of the components are outlined. The mathematical equations describing the various transmission effects are given in Chap. 3.

2.1 Generic Setup

The generic setup of a typical optical transmission system is shown in Fig. 2.1, and it is described briefly in the following. In this book long-haul transmission systems with a typical (transparent) reach of more than 1,000 km are investigated. In these systems single mode fibers are currently the transmission medium of choice.

Such optical fibers have a total usable bandwidth of several terahertz. This is why wavelength division multiplexing (WDM) is commonly employed to utilize this enormous capacity. A typical transmission system consists of an array of lasers with different wavelengths to generate the optical carriers. Each laser is modulated by an external modulator (e.g. a Mach–Zehnder modulator) to impress the data signal. Most commercially available transmission systems have a channel bit rate of 43 Gb/s (e.g. [1, 2]), and the next generation of optical transmission system operating at a line rate of 112 Gb/s (including FEC and Ethernet protocol overheads) is already available from some system vendors. As modulation format initally non-return-to-zero (NRZ) on–off keying (OOK) has been employed for line rates up to 10.7 Gb/s [3]. The following first generation of 43 Gb/s (per channel) transmission systems has been designed to use

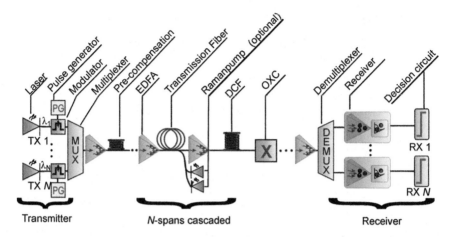

Fig. 2.1 Generic setup of a fiber optical wavelength division multiplexing transmission system

return-to-zero (RZ) OOK. Soon a duobinary implementation followed due to the higher tolerance to accumulated chromatic dispersion. Further improvements led to the use of differential phase shift keying (DPSK) with interferometric detection and to the current version using quadrature phase shift keying (QPSK) and polarization multiplexing (polmux) with coherent reception. QPSK polmux transmits a QPSK signal on each of the two orthogonal polarization axes, thus doubling the spectral efficiency. For long distance 112 Gb/s line rate systems coherent QPSK polmux transmission is widely adopted as the modulation format of choice [4].

To combine the different wavelengths for transmission on a single mode fiber a multiplexer (typically an arrayed waveguide grating, AWG, with fixed channel spacing) is employed [5]. After the multiplexer an optical amplifier (erbium doped fiber amplifier, EDFA) is deployed to increase the signal launch power. Potentially a dispersion pre-compensation fiber may be used. To improve the performance—especially in long fiber links—contra-directional (in some cases combined with co-directional) Raman pumping may be utilized in the transmission fiber. Along the transmission line the signal is periodically amplified (approximately every 80–100 km). In most of today's systems the accumulated group-velocity dispersion (GVD) is compensated after each span by a lumped dispersion compensating fiber (DCF). An alternative to DCFs are chirped fiber Bragg gratings (FBG). They have the advantage of a lower insertion loss which is also independent of the amount of dispersion to be compensated. A major disadvantage is the inherent phase ripple originating from the production process, which can lead to significant signal distortions, if several FBGs are concatenated [6].

At a node, where several different routes interconnect, an optical cross-connect (OXC) may be deployed. An OXC allows to switch from one fiber to another dynamically. It is also possible to route individual wavelengths in different directions [5].

At the receiver, the different wavelengths need to be demultiplex ed. This can be achieved by optical band pass filters (e.g. dielectric multicavity thin-film filters [7]) or an AWG. For the widely employed NRZ-OOK format the received signal is directly detected by a photo diode, which generates the electrical base band signal. Finally, the binary data stream is restored by a decision circuit. Novel modulation formats with coherent detection make use of an additional laser at the receiver to generate a local oscillator signal. This signal is mixed with the incoming signal using a so called 90°-hybrid. In polmux systems additionally a separation of the two polarization axes is needed, which is usually performed by a polarization splitter. After conversion into the electrical domain typically an electrical equalizer is employed (compare also Sect. 2.8). The equalizer is used to compensate for distortions along the transmission line (especially linear effects such as accumulated dispersion and polarization mode dispersion). Additionally the mismatch in phase and frequency between the transmitter laser and the local oscillator at the receiver must be compensated. This is also achieved by digital signal processors.

In the following paragraphs the various components mentioned above (compare also Fig. 2.1) are explained in more detail.

2.2 Transmitters

In fiber optical transmission systems transmitters consist of a light source used as the optical carrier and a modulator to impress the data signal onto this carrier.

In optical long-haul transmission networks mainly coherent continuous-wave (CW) lasers are used. Semiconductor lasers are by far the most popular light source for optical communication systems [5]. Semiconductor lasers are compact and usually only a few hundred micrometers in size. As they are essentially based on pn-junctions they can be fabricated in large volumes using highly advanced integrated semiconductor technology. Frequently used are DFB (distributed feedback) lasers, which are made of InGaAsP (indium gallium arsenide phosphide) for the required wavelength range. Today (wavelength) tunable lasers are highly desirable to relax inventory and sparing issues, which are rather expensive. They are also one of the key enablers of reconfigurable optical networks as they allow to choose the transmit wavelength arbitrarily at the source of a lightpath. DFB lasers can be tuned by varying the forward-bias current, which changes the refractive index. However, changing the bias current also changes the output power of the device and makes this technique unsuitable. This is why distributed Bragg reflector (DBR) lasers are typically used, if tuning is desired. In DBR lasers a separate current in the Bragg region can be used to tune the wavelength.

Lasers are usually not directly modulated to suppress the chirping of the emission wavelength, which is especially detrimental at higher bit rates (>2.5 Gb/s) [5]. Instead, external modulators are used e.g. optical Mach–Zehnder modulators

Fig. 2.2 Transmitter setup with an optical IQ modulator

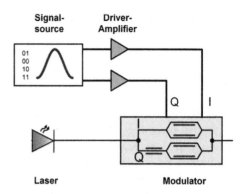

(MZM). MZMs can be implemented in Lithium Niobate (LiNbO$_3$), Gallium Arsenide (GaAs) or Indium Phosphide (InP). These modulators and the underlying electronics allow to directly generate signals of up to 40 GSamples/s. Field experiments have already shown 107 Gb/s ETDM transmission with NRZ-OOK directly driving an MZM [8]. Signals with significantly higher data rates (up to more than 1 Tb/s) are usually generated by optical time domain multiplexing (OTDM). Different optical signals of lower bit rates are interleaved in this process to generate a data signal of the desired higher modulation frequency.

If QPSK or even higher order modulation formats are to be transmitted, the modulator setup is getting more complicated [9]. In Fig. 2.2 the setup of a QPSK transmitter based on an optical IQ modulator is shown. Typically a nested Mach–Zehnder structure (IQ modulator) is used with a 90°-phase shift between the upper and lower branches allowing to modulate the real and imaginary parts of the signal independently. Such a device is available in an integrated form. For polmux transmission two of these IQ modulators are used. In this case the orthogonal carriers are provided from a laser source, which is split into two orthogonal polarizations by a polarization beam splitter (PBS).

As already mentioned above, the high bandwidth of the fiber can be utilized efficiently by employing a multitude of transmitters using different wavelengths. This transmission scheme is called wavelength division multiplexing (WDM). With the help of a multiplexer the different wavelengths are combined and transmitted over the same transmission fiber. Often an arrayed waveguide grating (AWG) is utilized as a multiplexer [5]. An AWG is a generalization of the Mach–Zehnder interferometer. It consists of two multiport couplers interconnected by an array of waveguides. In this array several copies of the same signal, but shifted in phase by different amounts, are superimposed. The AWG can be used as an $n \times 1$ wavelength multiplexer as well as a $1 \times n$ wavelength demultiplexer. The main drawback of AWGs is their relatively high cost and lack of scalability.

In todays commercially available transmission (e.g. [1, 2]) systems up to 160 channels can be transmitted simultaneously.

2.3 Modulation Formats

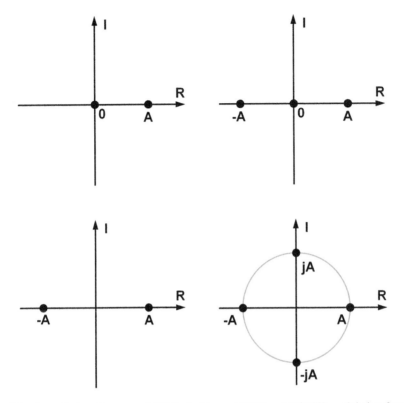

Fig. 2.3 Constellation diagrams of OOK, duobinary, DPSK and DQPSK modulation formats (from *left* to *right* and *top* to *bottom*)

2.3 Modulation Formats

The most commonly used modulation scheme in optical communications has been on–off keying (OOK) over years [10]. This means that the optical power is modulated according to the binary data input signal. Two main types of line codes are usually employed: return to zero (RZ) and non-return to zero (NRZ) formats. In RZ formats, the signal amplitude returns to zero at the boundaries of each bit slot, even if consecutive marks are sent whereas in NRZ format the signal amplitude remains on the high level in the case of consecutive marks.

The input data signal may be additionally pre-processed to generate more complex signal forms such as duobinary [9]. The duobinary modulation format is a pseudo three level modulation format. Apart from the mark signal in the unipolar complex plane also a signal with 180° phase shift is admissible (leading to the symbols $-1, 0, 1$ also shown in Fig. 2.3, top right).

Duobinary modulation is a special partial-response code. It can be generated from a binary NRZ signal by delay-and-add coding or low pass filtering (Fig. 2.4).

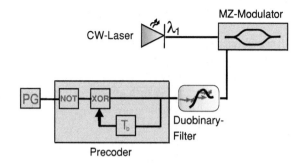

Fig. 2.4 Duobinary transmitter with single-arm MZ modulator. This setup avoids symmetry requirements of the MZ modulator

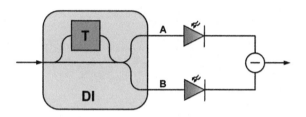

Fig. 2.5 Schematic setup of a balanced receiver using an optical delay line interferometer

The receiver setup can remain unmodified compared to standard direct detection systems. The main advantages of duobinary modulation are that the signal is more robust against accumulated chromatic dispersion induced signal distortions and that it shows a narrower spectrum compared to NRZ-OOK. Furthermore, a more complex setup is needed only at the transmitter. This is why duobinary modulation has been deployed in some 43 Gb/s transmission systems.

Recently, phase modulation formats and higher order modulation have become more widely studied [9, 10]. At first the focus has been primarily on differential phase shift keying (DPSK). DPSK encodes information on the binary phase change between adjacent bits. A mark is encoded onto a π phase change, whereas a zero is represented by the absence of a phase change. The main advantage compared to NRZ-OOK is a 3 dB receiver sensitivity improvement, which can be intuitively understood from the increased symbol spacing for DPSK compared to OOK for fixed average optical power [9]. This signal format furthermore shows better behavior regarding signal distortions due to fiber nonlinearities especially excellent resilience to cross-phase modulation, which is a key requirement for DWDM system implementation. DPSK, however, cannot be directly received using square-law detection of a photodiode. This is why—if direct detection is desired—a delay line interferometer (DI) is inserted in the optical path at the receiver to convert the differential phase modulation into intensity modulation (Fig. 2.5). DPSK has also been employed in some commercial 43 Gb/s transmission systems (e.g. [1]).

Currently, multilevel modulation formats (with a capacity of more than one bit per symbol) are state-of-the-art. Especially DQPSK has been widely studied (e.g. [9, 11, 12]). DQPSK transmits the four phase shifts 0, $+\pi/2$, $-\pi/2$, π at a symbol

2.3 Modulation Formats

rate of half the bit rate. As already mentioned above DQPSK is most conveniently implemented by two nested MZMs. DQPSK has the advantage that the spectrum is compressed in frequency by a factor of two compared to DPSK. This is beneficial for achieving a higher spectral efficiency in WDM systems, but also for increased tolerance to chromatic dispersion as well as a higher robustness concerning polarization mode dispersion (PMD). At the receiver the DQPSK signal can be detected by two balanced receivers using optical DIs.

The most recent version of optical long-haul transmission systems uses 43 Gb/s coherent polmux QPSK transmission, and this modulation format is also widely chosen for next generation optical transmission systems with line rates of 112 Gb/s [4]. Compared to the previously outlined modulation formats, which use only a single polarization axis, the spectral efficiency is doubled in polmux transmission by utilizing two orthogonal polarization axes. To enable access to both polarization axes the incoming signal is split into its orthogonal parts by a polarization splitter before the receiver. These two signals are coherently detected by mixing with a local oscillator in so called 90°-hybrids. This makes available both amplitude and phase information. The detailed configuration will be shown in Sect. 2.7.

In contrast to all previously described modulation formats, in coherent polmux transmission, digital signal processing is used extensively at the receiver (compare also Sect. 2.8). This requires information on both the amplitude and the phase of the incoming signals. Both are only available, if coherent detection is employed. Because it is hard to design an optical phase-locked loop the difference between the local oscillators at the receiver and the transmitter with respect to (carrier) frequency and phase have to be compensated in the electrical domain. Furthermore, the digital signal processing must compensate for the mixing of the orthogonal polarization components induced by PMD. As a benefit (linear) degradation effects can be compensated more easily by electrical equalization.

For the upcoming generation of optical transmission systems higher-order modulation formats attract increasing interest [10, 13]. 16-QAM modulation may become the format of choice for transmission systems with 224 Gb/s (compare e.g. [13]). For even higher bit rates more sophisticated higher level modulation formats may be used in the future.

Also orthogonal frequency division multiplexing (OFDM) has been widely studied in the recent years. OFDM has the advantage of being very tolerant to linear signal distortions such as chromatic dispersion and PMD. Furthermore, in combination with coherent detection electronic equalization can be implemented easily. Also adaptation to the desired reach is possible because the signal constellation and use of subcarriers can be changed electronically. On the other hand OFDM requires a relatively high resolution of DAC and especially ADC converters, which may be difficult to realize at the required high bandwidth [14]. So far, however, no commercial realization for long-haul transmission at high bitrates is foreseeable. A short introduction into optical OFDM is given in e.g. [15]. A good overview and comparison between OFDM and QPSK modulation can be found in e.g. [16].

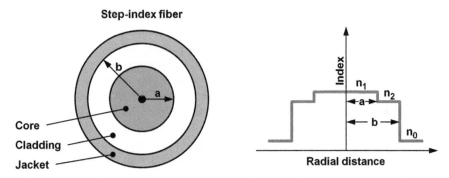

Fig. 2.6 Cross-section and refractive index profile of a single mode step-index fiber (adapted from [18])

2.4 Fiber Properties

In today's long and ultra-long haul networks solely single mode fibers (SMF) are used [17]. These fibers guide a single transmission mode only (but two orthogonal polarizations). Single mode glass fibers consist of a cylindrical core of silica glass surrounded by a cladding whose refractive index is lower than that of the core (Fig. 2.6, left). As the refractive index changes abruptly from the cladding to the core (Fig. 2.6, right) they are called step-index fibers. The diameter of a single mode fiber core ($2 \cdot a$) is approximately 8–10 μm and the cladding diameter ($2 \cdot b$) is typically 125 μm.

In Fig. 2.7 the spectral characteristics of dispersion and attenuation are plotted. There are two operational windows, which are normally used for optical communications. The windows are centered on 1,550 nm (usually used for long-haul communications) and 1,300 nm (usually used for metro networks) and are separated by OH^- absorption peaks. In today's transmission systems, usually the C- and L-bands (conventional- and long-band, compare Fig. 2.7) are used, which allow a total number of up to 160 channels using a channel spacing of 0.4 nm (50 GHz) in dense wavelength division multiplexing (DWDM) configuration. To utilize an even larger spectral bandwidth, new fiber types have been developed, which do not show the OH^- absorption peak and are usually referred to as "AllWave" fibers.

The reasons for using the C- (1,530–1,565 nm) and L-bands (1,565–1,625 nm) are the low attenuation of the transmission fiber in this wavelength region and the availability of optical amplifiers, so called Erbium-doped fiber amplifiers (EDFA, also compare the following section). In combination with gain flattening filters, EDFAs offer a flat gain spectrum over the whole C- and L-bands. Additionally the S- and U-bands may be used in future optical networks. In this case, a combination of Raman amplification, which is available in all bands and other lumped rare-earth doped amplifiers, may be employed.

Apart from the attenuation of the fiber, the spectral characteristics of the dispersion are important. These characteristics are measured by the dispersion

2.4 Fiber Properties

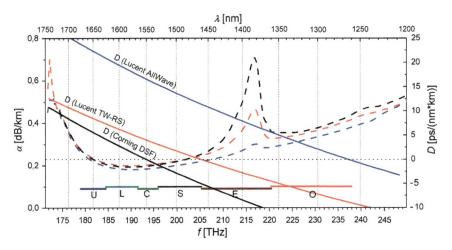

Fig. 2.7 Spectral characteristics of the dispersion (*solid lines, right axis*) and attenuation (*dashed lines, left axis*) as well as transmission band nomenclature

parameter D [ps/(nm · km)]. From the dispersion point of view four different fiber types can be distinguished: standard single mode fibers (SSMF), non-zero dispersion shifted fibers (NZDSF), dispersion shifted fibers (DSF) and dispersion compensating fibers (DCF). An overview of the fiber parameters is given in Table A1 in the appendix. The SSMF has $D \approx 0$ at 1,300 nm and a dispersion value of $D \approx 17$ ps/(nm · km) in the C-band. This is why (accumulated) dispersion compensation is needed, which can be done optically e.g. by the use of DCF. DCFs have a high negative dispersion value in the C-band. In contrast to that DSFs have $D \approx 0$ in the C-band. This offers the advantage that no dispersion compensation is needed. In WDM systems, though, large impairments occur due to nearly perfect phase matching of the different channels, which leads to strongly increased nonlinear effects (especially four-wave mixing) in the fiber. Consequently, NZDSF fibers have been developed, which attempt to create a compromise between both fiber types. They offer a relatively low local dispersion of $D \approx 4$ ps/(nm · km) in the C-band. Thus they suppress the nonlinear effects more effectively than DSFs, and they offer a performance advantage over SSMFs in terms of less accumulated dispersion.

Furthermore, the dispersion slope parameter S [ps/(nm² · km)] is used to model the spectral tilt of the dispersion. It is extremely important to additionally compensate for the slope of the dispersion, if DWDM systems with a high channel count are used, which cover a larger bandwidth and operate at high channel bit rates ($\geq$10Gb/s). Otherwise, the edge channels of the spectrum may still encounter a considerable amount of degradation due to uncompensated dispersion stemming from imperfect matching of transmission fiber and DCF slope values. In high bit rate systems of 40 Gb/s or more also the temperature dependence of chromatic dispersion becomes noticeable. The fiber cables are normally buried 0.6–1.2 m

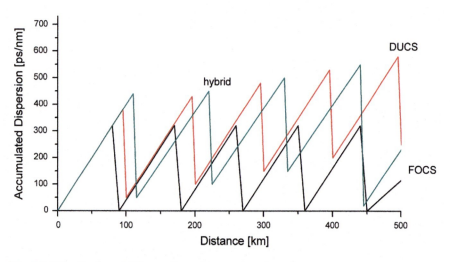

Fig. 2.8 Visualization of different dispersion compensation schemes

below the surface. In this depth the cable temperature can vary by several degrees over the year [19].

For the suppression of the nonlinear effects, the dispersion compensation scheme plays an important role. Several different dispersion compensation schemes exist in today's transmission systems.

First, there is the full-inline optimized post-compensation scheme (FOCS), where the accumulated dispersion is returned to (approximately) zero after each span. A major disadvantage of FOCS is that after each span the same phase conditions are restored leading to a strong accumulation of especially nonlinear fiber effects. This is why the distributed under-compensation scheme (DUCS) is commonly employed. In this scheme after each span some residual dispersion remains, which is advantageous for suppressing nonlinear fiber effects. Recently, a combination of DUCS and selective over-compensation has been suggested (hybrid dispersion compensation, compare Fig. 2.8). Hybrid compensation offers the advantage of some residual dispersion after each span while not accumulating a very high amount of residual dispersion, which needs to be compensated in front of the receiver.

Also pre-compensation is likely to be employed in WDM transmission systems with NRZ-OOK channels. Dispersion pre-compensation means that a certain amount of dispersion is compensated already at the beginning of the transmission system before the first transmission fiber. More information can be found in e.g. [20], where also design rules for the right amount of pre-compensation have been presented.

In today's high data rate transmission systems using coherent detection, the dispersion can be compensated electronically on a channel-by-channel basis to obtain the maximal system performance making it possible to build transmission systems without in-line dispersion compensation. However, as often legacy systems are upgraded, most systems are very likely to contain inline DCF modules for a long time in the future.

2.5 Amplifiers   21

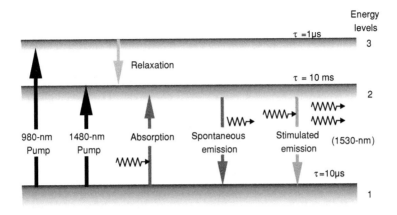

Fig. 2.9 Visualization of EDFA energy levels and working principle (adapted from [21])

2.5 Amplifiers

As in an optical communication system the optical signal is attenuated by the optical fiber and other components such as multiplexers and couplers, it has to be reamplified after some distance. Instead also so-called regenerators may be deployed, which receive the signal and retransmit it again. Today mostly optical amplification with Erbium-doped fiber amplifiers (EDFA) is used as it allows amplifying the entire WDM spectrum simultaneously [21]. Usually every 80–120 km an EDFA is deployed. The Erbium-doped fiber (EDF) in these amplifiers has a length of a few meters. In the EDF electrons are raised to a higher energy level by optical pumping with a lower wavelength laser (compare Fig. 2.9). The electrons fall back to their original level, stimulated by incoming signal photons. During this process, they emit a duplicate of the original photon. EDFAs work similar to lasers but without a feedback cavity. However, apart from the stimulated emission of photons there is also spontaneous emission, which adds noise to the signal. This noise is referred to as amplified spontaneous emission noise (ASE) because the photons generated by spontaneous emission are amplified along the rest of the EDF. In modern WDM systems EDFAs with several stages are widely used. The first stage of the EDFA has a low noise figure because it needs to provide a low output power only. The second stage of the EDFA, the booster, is used afterwards to generate a high output power signal. The noise performance of the whole amplifier is dominated by the first stage. Thus the 2-stage amplification produces a high-performance amplifier with low noise and high output power. Between the two amplifier stages a loss element can be placed with negligible impact on the performance. Typically a DCF module is connected at this point, and also a gain flattening filter is inserted [5].

An alternative to lumped amplification using EDFAs is distributed amplification using the Raman effect [22]. To utilize this nonlinear effect, Raman pumps with lower wavelengths than the signal are fed into the fiber. Normally, contra-directional pumping is employed. As a result of the Raman effect, power is transferred from the lower wavelengths to the higher ones. The efficiency of Raman amplification (on standard transmission fibers and limited pump powers) is significantly lower than the gain, which can be obtained by EDFAs. That is why mostly a combination of Raman amplification and EDFAs is used. The greatest advantage of Raman amplifiers is that they work in every waveband. Furthermore, the OSNR is improved by Raman amplification because the minimum signal power along the transmission line is increased. Today the most popular use of Raman amplifiers is to complement EDFAs by providing additional gain in a distributed manner in long-haul transmission systems with some extra long fiber spans.

2.6 Optical Cross Connects

To enable dynamic reconfiguration of the transmission system along the transmission path optical cross connects (OXCs) may be deployed. In the literature OXCs exist in opaque and transparent configurations. In this book OXCs are used only in the all-optical configuration leaving the signal in the optical domain as it passes through the node.

Transparent OXCs are replacements for manual fiber patch panels. There are two main applications: protection switching to another lightpath in the case of a network failure (or planned maintenance) and provisioning of new lightpaths. Many optical technologies are available to realize optical switches [5]. OXCs can be based on micro-electro-mechanical systems (MEMS), on thermo-optical silica based systems, on electro-optics lithium niobate systems or on liquid crystals to enumerate only a few possible realizations.

The simplest realization of a MEMS based switch uses 2D mirrors. In one state, the mirror remains flat in line with the substrate and in the other state the mirror pops up to a vertical position and the light beam is deflected (Fig. 2.10).

Thermo-optical silica based switches are essentially 2×2 integrated-optic Mach–Zehnder interferometers. By varying the refractive index in one arm, the relative phase difference between the two arms is changed, leading to a switching from one output port to the other. Electro-optical switches can also be built in the Mach-Zehnder configuration, however, this time the electro-optical effect is used to change the relative phase.

Liquid crystals allow rotating the polarization of the incoming light based on an applied external voltage. In a configuration with a polarization beam splitter and a polarization beam combiner in this way the signal can be switched from one port to the other (for more details see e.g. [5]). Furthermore, the liquid crystal can be used as a variable optical attenuator (VOA) to control the output power of the signal.

To allow switching on a wavelength granularity level a (wavelength) demultiplexer is deployed in front of the OXC to split the incoming signal in its individual wavelengths. Afterwards the same switch architectures as described above

2.6 Optical Cross Connects

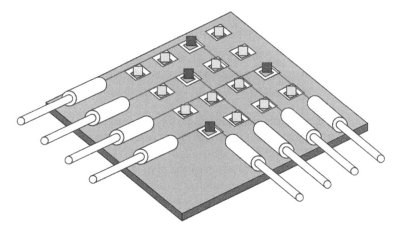

Fig. 2.10 4 × 4 MEMS matrix switch using free-space paths and 2D- movable mirror elements [23]

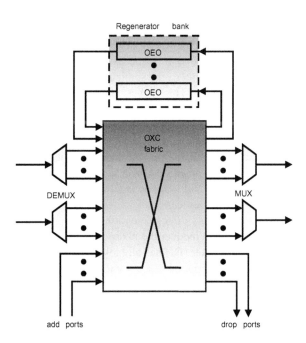

Fig. 2.11 Optical cross connect with optional optical electrical-optical regeneration

may be utilized. On the output of the OXC a multiplexer is used to combine the different wavelengths again.

Optical cross connects can also be combined with an optical-electrical-optical (OEO) regenerator bank (Fig. 2.11). In this way some channels, which are highly degraded e.g. because of a long transparent path length, may be passed to a regenerator to recover the signal.

2.7 Receivers

At the receiver the optical WDM signal is demultiplexed. For this purpose an identical arrayed waveguide grating (AWG) may be used as for the multiplexer. After demultiplexing the signal has to be converted from the optical to the electrical domain.

In the case of intensity modulated signals, the power of each WDM channel can be directly fed to a photodiode (direct detection), which converts the optical signal into an electrical one. Direct detection can only recover the intensity of the electric field due to the square transfer function of the photodiode. This has the advantage that no phase-, frequency- or polarization control is necessary. On the contrary the information encoded in the optical phase can only be obtained when employing additional components (e.g. delay line interferometers).

Photodetectors are made of semiconductor materials (more details can be found in e.g. [5, 24]). Photons incident on such a semiconductor are absorbed by electrons in the valence band. As a result, these electrons acquire higher energy and are excited in the conduction band (if the band gap is chosen correctly), leaving behind a "hole" in the valence band. When an external voltage is applied to the semiconductor, these electron–hole pairs give rise to an electrical current (the photocurrent). In practice it is necessary to sweep the generated conduction band electrons rapidly out of the semiconductor before they can recombine with holes in the valence band. This is best achieved by using a semiconductor pn-junction instead of a homogeneous slab semiconductor. To improve the efficiency of the photodiode, a very lightly doped intrinsic (i-type) semiconductor is introduced between the p-type and n-type semiconductors. Such a photodiode is called pin photodiode. Typically pin photodiodes for the C- and L-bands are built from indium gallium arsenide (InGaAs) for the intrinsic zone and indium phosphide (InP) for the p- and n-zones (so called double heterostructures).

As photodiodes can only detect the optical intensity, in the case of phase modulated signals direct detection is not possible. A different receiver structure has to be used to recover the phase information. A possible realization with a delay line interferometer (DI) has already been depicted in Fig. 2.5. A DI can be used to convert the differential phase modulation into intensity modulation.

If in the electrical domain both amplitude and phase information are needed (e.g. because digital signal processing and equalization are desired), coherent reception must be employed. The setup of a coherent receiver for a polmux QPSK system is shown in Fig. 2.12. A local oscillator (LO) laser is used to provide a local reference signal. Two fundamental principles are distinguished: homodyne and heterodyne detection. In the former case of homodyne detection, the carrier frequencies of the LO and the signal laser are (almost) identical and the optical data signal is directly converted to the electrical baseband signal. One of the main challenges, though, is to synchronize the carrier frequencies and phases of the signal laser at the transmitter and the LO at the receiver. In the case of heterodyne detection, the frequencies of the signal laser and the LO are chosen to be different

2.7 Receivers

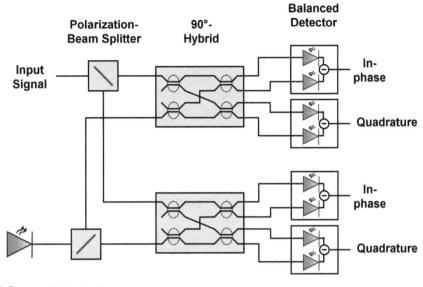

Fig. 2.12 Setup of coherent receiver for polmux QPSK signals

leading to an electrical signal at an intermediate frequency (IF). The advantage of heterodyne detection is that it permits simpler demodulation schemes and enables carrier synchronization with an electrical phase locked loop. On the other hand, the occupied electrical bandwidth is more than twice as high as for homodyne detection.

The mixing of the LO laser and the input signal is usually achieved by a so-called 90°-hybrid. It enables detection of the in-phase and quadrature components of an optical signal. Different realization options can be found in [25].

In principle single-ended detection by a photodiode would be possible at the upper and lower outputs of the 90°-hybrid. To not waste half of the power, however, a balanced detector is usually applied and the difference of both photodiodes is passed onwards.

For polmux operation additional care has to be taken to the demultiplexing of the two orthogonal polarization components. For this purpose polarization beam splitters are used to split up the LO laser and the input signal into parallel and orthogonal components. For each of these two polarization components an identical setup of 90°-hybrid and balanced detector is used leading to four output signals (two in-phase and two quadrature signals). In the electrical domain the four outputs are analog-to-digital converted. Afterwards, polarization demultiplexing and compensation of degradation effects is accomplished by adaptive digital signal processing.

2.8 Electrical Signal Processing

Digital (electrical) signal processing (DSP) has been emerging as a practical solution for long-haul optical communications for some years (compare e.g. [26 or 27]). Historically optical communications have operated at the very limits of electronic technology, preventing the application of digital signal processing which has become prevalent in many other fields such as e.g. wireless communications.

The first generation of electronic signal processing in optical communications emerged in the late 1980s and introduced forward error correction (FEC) to the digital (binary) data sequence. The first FEC codes (in optical communications) have been applied in 1988 by Grover [28]. Over the next years FEC has been standardized for use in optical transmission by the ITU-T with a 7% out-of-band overhead. Current codes (e.g. RS(1901, 1855)) yield a net coding gain of 8.7 dB at an output BER $= 10^{-13}$.

The second generation of electronic signal processing used some limited analog functions such as feed-forward equalizers (FFE) or decision directed equalizers (DDE). As in these days direct detection has been prevalent, the signal processing has been very limited due to the loss of phase information. In the third generation the main improvement has been the introduction of analog-to-digital converters (ADCs). DSPs replaced the analog signal processing and the clock and data recovery circuitry [26]. Maximum likelihood sequence estimation (MLSE) has been studied intensively in this period (compare e.g. [29]).

In the fourth generation electronic pre-distortion (EPD) of the signal at the transmitter side has been proposed. In contrast to the MLSE approach the complexity has been reduced and allowed precompensation of the chromatic dispersion and also nonlinear (single channel) effects (compare e.g. [30]). The main drawback of such a system is the limited robustness to interchannel crosstalk stemming from nonlinear effects such as four-wave mixing (FWM) and especially cross-phase modulation (XPM) and the limited flexibility inherent with predistortion (e.g. exact knowledge of the length of the transmission line and the dispersion parameters is required).

Current DSP systems operate on the electrical signals obtained from the outputs of a coherent receiver as shown in Fig. 2.12 and are implemented in application specific integrated circuits (ASICs). DSP are used for the compensation of (linear) transmission impairments and polarization demultiplexing [31]. One of the main challenges is that the algorithms have to be as simple as possible to enable high-speed real time processing. The adaptation speed is usually of less concern as the fiber channel varies rather slowly (compared to e.g. wireless communication systems) making blind adaptation attractive. The first step in the DSP is usually the compensation of the accumulated chromatic dispersion (CD) using e.g. a finite impulse response (FIR) filter structure. Another possibility is the implementation with infinite impulse response (IIR) filters, where the required tap number is lower than for FIR filters. However, the feedback of the IIR equalizers makes this approach hard to implement in high-speed electronics with parallelized signal processing [27]. Ideally the dispersion estimation can be performed automatically allowing an optimum channel-by-channel compensation of the actual accumulated dispersion. An example of a possible realization is given in [32]. The next task is timing recovery to correct for the timing phase and frequency offset between the transmitter and receiver clocks. The phase error can be derived e.g. with the help of the Gardner algorithm [33]. The correction of the timing phase and frequency is performed either in the digital domain with the help of an interpolator or directly in an ADC. Afterwards polarization demultiplexing is performed. For this problem the fiber optic channel can be seen as a special case of the more general multiple-input-multiple-output (MIMO) systems. The constant modulus algorithm (CMA) is a popular choice for the approximate acquisition [34]. It is often followed by a least-mean squares (LMS) algorithm for tracking purposes. However, the LMS is decision directed requiring prior correction of the carrier phase and frequency offset. These offsets stem from the mixing with the local oscillator and are identical in both polarizations. The consequence is a rotating constellation diagram. A typical implementation of the phase offset estimation is based on the Viterbi-and-Viterbi algorithm [35]. The LO-phase recovery should have a rather low bandwidth in order to minimize the noise influence which is even worsened by the rapid XPM-induced phase fluctuations. Further improvements to the standard implementation using the Viterbi-and-Viterbi algorithm processing the two polarization planes independently can be gained by a joint-polarization (JP) algorithm [36]. These improvements lead to a significant increase of the ASE-noise tolerance.

Concluding one can say that digital signal processing has become a central element in state-of-the-art coherent optical communication systems. It will be interesting to see whether electrical signal processing will keep pace with the increasing data rates of optical communications and even more complex signal processing will be an integral part of future transmission systems.

References

1. Nokia Siemens Networks. SURPASS hiT 7500—Brochure, http://www.nokiasiemensnetworks.com/sites/default/files/document/hiT_7500_0.pdf
2. Alcatel-Lucent. 1625 LambdaXtreme Transport—Brochure, Release 8.1 (2009)
3. Gnauck, A.H., Tkach, R.W., Chraplyvy, A.R., Li, T.: High-capacity optical transmission systems. IEEE/OSA J. Lightw. Technol. 26(9), 1032–1045 (2008). May
4. Birk, M., Gerard, P., Curto, R., Nelson, L., Zhou, X., Magill, P., Schmidt, T.J., Malouin, C., Zhang, B., Ibragimov, E., Khatana, S., Glavanovic, M., Lofland, R., Marcoccia, R., Saunders, R., Nicholl, G., Nowell, M., Forghieri, F.: Coherent 100 Gb/s PM-QPSK field trial. IEEE Comm. Mag. 48(7), 52–60 (2010). July
5. Ramaswami, R., Sivarajan, K.N., Sasaki, G.H.: Optical networks—a practical perspective, 3rd edn. Morgan Kaufmann, Burlington (2010)
6. Kaminow, I., Li, T.: Optical fiber an telecommunications IV B. Academic Press, San Diego (2002)
7. Vossen, J.L., Kern, W.: Thin Film Processes. Academic Press, NY (1978)

8. Jansen, S. L., Derksen, R. H., Schubert, C., Zhou, X., Birk, M., Weiske, C. -J., Bohn, M., van den Borne, D., Krummrich, P. M., Möller, M., Horst, F., Offrein, B. J., de Waardt H., Khoe, G. D., Kirstädter, A., 107-Gb/s full-ETDM transmission over field installed fiber using vestigial sideband modulation, Optical Fiber Communication Conference (OFC), OWE3, March 2007
9. Winzer, P.J., Essiambre, R.-J.: Advanced modulation formats for high-capacity optical transport networks. IEEE/OSA J Lightw. Technol. **24**(12), 4711–4728 (December 2006)
10. Seimetz, M.: Higher-order modulation for optical fiber transmission. Springer, Berlin (2009)
11. Griffin, R. A., Carter, A. C.: Optical differential quadrature phase shift key (oOQPSK) for high capacity optical transmission. Optical Fiber Communication Conference (OFC), WX6, March 2002
12. Griffin, R. A., Johnstone, R. I., Walker, R. G., Hall, J., Wadsworth, S. D., Berry, K., Carter, A. C., Wale, M. J., Hughes, J., Jerram, P. A., Parsons, N. J.: 10 Gb/s optical differential quadrature phase shift key (DQPSK) transmission using GaAs/AlGaAs integration, Optical Fiber Communication Conference (OFC), FD6, March 2002
13. Alfiad, M., Kuschnerov, M., Jansen, S., Wuth, T., van den Borne, D., de Waardt, H.: Transmission of 11 x 224 Gb/s POLMUX-RZ-16QAM over 1500 km of LongLine and pure-silica SMF, European Conference on Optical Communication (ECOC), We.8.C.2, Torino, September 2010
14. Pachnicke, S., Özdür, S., Griesser, H., Fürst, C., Krummrich, P.: Sensitivity to signal quantization of 43 Gb/s and 107 Gb/s optical 16-QAM OFDM transmission with coherent detection. Elsevier J. Opt. Fiber Technol. **15**(5), 414–419 (2009)
15. Lowery, A.: Adaptation of orthogonal frequency division multiplexing (OFDM) to compensate impairments in optical transmission systems, European Conference on Optical Communication (ECOC). Tutorial, Berlin, (2007)
16. Jansen, S.L., Spinnler, B., Morita, I., Randel, S., Tanaka, H.: 100GbE: QPSK versus OFDM. Elsevier J. Opt. Fiber Technol. **15**(5), 407–413 (2009)
17. Kao, K.C., Hockham, G.A.: Dielectric-fibre surface waveguides for optical frequencies. IEE Proc. **113**(7), 1151–1158 (1966)
18. Agrawal, G.P.: Fiber optic. comm. sys. Wiley, New York (1992)
19. Vorbeck, S., Leppla, R.: Dispersion and dispersion slope tolerance of 160-Gb/s systems, considering the temperature dependence of chromatic dispersion. IEEE Photon Technol. Lett. **15**(10), 1470–1472 (2003)
20. Lehmann, G., Meissner, E.: Dispersion management for the transmission of mixed data rates of 40 Gb/s and 10 Gb/s in the same fiber. Proc. SPIE **4906**, 253–260 (2002)
21. Desurvire, E.: Erbium doped fiber amplifiers–principles and applications. Wiley, New York (1994)
22. Islam, M.N.: Raman amplifiers for telecommunications 2: sub-systems and systems. Springer, Berlin (2003)
23. Hoffmann, M.: Micro-opto-electro-mechanical systems on silicon for optical communications systems. Habilitation, Dortmund (2002)
24. Voges, E., Petermann, K.: Optische Kommunikationstechnik. Springer, Berlin (2002)
25. Seimetz, M., Weinert, C.M.: Options, feasibility and availability of 2 x 4 90°-hybrids for coherent optical systems. IEEE/OSA J. Lightw. Technol. **24**(3), 1317–1322 (2006)
26. Savory, S.J.: Electronic signal processing in optical communications. In: Proc. of SPIE Vol. 7136, SPIE Asia-Pacific Optical Communications Conference (APOC 2008). Hangzhou, China (2008)
27. Kuschnerov, M., Hauske, F.N., Piyawanno, K., Spinnler, B., Alfiad, M.S., Napoli, A., Lankl, B.: DSP for coherent single-carrier receivers. IEEE/OSA J. Lightw. Technol. **27**(16), 3614–3622 (2009)
28. Grover, W.D.: Forward error correction in dispersion-limited lightwave systems. IEEE/OSA J. Lightw. Technol. **6**(5), 643–645 (1988)

29. Kupfer, T., Dorschky, C., Ene, M., Langenbach, S.: Measurement of the performance of 16-States MLSE digital equalizer with different optical modulation formats, Optical Fiber Communication Conference (OFC), PDP13, Anaheim, USA, March 2008
30. Killey, R. I., Watts, P. M., Glick, M., Bayvel, P.: Electronic dispersion compensation by signal predistortion. In: Optical Fiber Communication Conference (OFC), OWB3, Anaheim, USA, March 2006
31. Fludger, C.R.S., Duthel, T., van den Borne, D., Schulien, C., Schmidt, E.-D., Wuth, T., Geyer, J., De Man, E., Khoe, G.-D., de Waardt, H.: Coherent equalization and POLMUX-RZ-DQPSK for robust 100-GE transmission. IEEE/OSA J. Lightw. Technol. **26**(1), 64–72 (2008)
32. Kuschnerov, M., Hauske, F. N., Piyawanno, K., Spinnler, B., Napoli, A., Lankl, B.: Adaptive chromatic dispersion equalization for no-dispersion managed coherent systems. In: Optical Fiber Communication Conference (OFC), OMT1, San Diego, USA, March 2009
33. Gardner, F.M.: A BPSK/QPSK timing-error detector for sampled receivers. IEEE Trans. Commun. **34**(5), 423–429 (1986)
34. Savory, S.J.: Digital filters for coherent optical receivers. OSA Opt. Exp. **16**(2), 804–817 (2008)
35. Viterbi, A.J., Viterbi, A.M.: Nonlinear estimation of PSK-modulated carrier phase with applications to burst digital transmission. IEEE Trans. Inf. Theory **29**(4), 543–551 (1983)
36. Kuschnerov, M., van den Borne, D., Piyawanno, K., Hauske, F. N., Fludger, C. R. S., Duthel, T., Wuth, T., Geyer, J. C., Schulien, C., Spinnler, B., Schmidt, E. -D., Lankl, B.: Joint-polarization carrier phase estimation for XPM-limited coherent polarization-multiplexed QPSK transmission with OOK-neighbors. In: European Conference on Optical Communication (ECOC), Mo.4.D.2, Brussels, September 2008

Chapter 3
Simulation of Fiber Optical Transmission Systems

Abstract This chapter deals with modeling and simulation of fiber optical transmission systems. In the first section the most basic properties of optical signal propagation through a fiber are presented in the form of mathematical equations. Afterwards the split-step Fourier method is explained, which is typically used to solve the nonlinear Schrödinger equation numerically. Furthermore, numerical and semi-analytical models to estimate the bit error ratio are described. At the end of this chapter the numerical photonic system simulation tool *PHOTOSS* is shown, which has been used for many simulations presented in this book.

3.1 Modeling of Fiber Optical Transmission Systems

In current long-haul optical transmission systems the modulation frequency (typical values lie in the range of approximately 2.5–100 GHz) is much lower than the optical carrier frequency (which is around 193.1 THz). This allows considering the modulated information (envelope) and the carrier optical field independently. Optical communication systems are usually modeled in the (complex) baseband. The baseband frequency spectrum of the optical signal is generated by shifting the modulated spectrum to around 0 Hz.

The electric field $E(t)$ of a modulated lightwave with a carrier frequency ω_0 can be written as:

$$E(t) = A(t)e^{j\omega_0 t} + A^*(t)e^{-j\omega_0 t} \qquad (3.1\text{-}1)$$

where $A(t)$ is the complex envelope of the lightwave, which is assumed to change much more slowly than the carrier ($\partial A/\partial t \ll f_c$). The spectrum of the base band signal is replicated at the positive and negative carrier frequencies. For an efficient simulation with a significantly lower number of required samples, the signal $E(t)$ is represented by the complex envelope $A(t)$. The spectra of $E(t)$ and $A(t)$ are related to each other by a simple frequency shift by the value of the carrier frequency.

The fiber is the key component in the simulation of optical communication systems. Most of the signal degradation acquired during transmission is a result of its physical properties. Therefore it must be modeled with great accuracy. Nevertheless, the efficiency of the simulation program—regarding computational time—is also important. The higher the desired accuracy, the higher the computational effort, which can lie in the range of days or even weeks for the simulation of a single long-haul transmission system on a state-of-the-art desktop PC (e.g. Intel® Core™ i7).

The propagation of optical signals in a fiber is modeled by the nonlinear Schrödinger equation (NLSE). The NLSE describes the effects of attenuation, dispersion, fiber nonlinearities as well as polarization-dependent effects and their interactions. For arbitrary input signals the NLSE can only be solved numerically. In most cases for this purpose the split-step Fourier method (SSFM) is used, which is described below. The effects stemming from fiber birefringence can be modeled using the so called wave plate model.

3.1.1 Attenuation and Dispersion

In the simplified case of a linear fiber, where all nonlinear effects occurring in a fiber have been set to zero or are assumed to be negligible, the propagation of the optical signal along the fiber—taking into account the effects of dispersion to the third order—can be modeled by the following differential equation [1]:

$$\frac{\partial}{\partial z}A(z,t) + \left(\frac{\alpha}{2} - j\beta_0 - \beta_1\frac{\partial}{\partial t} + \frac{j}{2}\beta_2\frac{\partial^2}{\partial t^2} - \frac{1}{6}\beta_3\frac{\partial^3}{\partial t^3}\right)A(z,t) = 0 \qquad (3.1\text{-}2)$$

where β_x refers to the propagation constant and its (x-th) derivatives to the angular frequency and stands for the fiber attenuation in Neper/km. The latter value can easily be obtained from the attenuation coefficient given in dB/km by

$$\alpha_{Neper/km} = \alpha_{dB/km} \cdot \frac{\ln 10}{10} \approx \alpha_{dB/km} \cdot 0.23. \qquad (3.1\text{-}3)$$

Two fundamental loss mechanisms govern the loss profile of an optical fiber: Rayleigh scattering and intrinsic absorption [2]. Rayleigh scattering results from local microscopic fluctuations in the material density that are created during the production process. The reason is that silica molecules move randomly in the molten state and freeze in place during fiber fabrication. The density fluctuations lead to small variations within the refractive index of the glass causing scattering. The scattering cross section varies with λ^{-4} making Rayleigh scattering the dominant loss mechanism for short wavelengths. For long wavelengths

3.1 Modeling of Fiber Optical Transmission Systems

(>1660 nm) intrinsic absorption becomes dominant. The reason for intrinsic absorption is the vibrational resonance associated with specific molecules. For silica molecules, electronic resonances occur in the ultraviolet region, whereas vibrational resonances occur in the infrared region. Because of the amorphous structure of silica, these resonances result in absorption bands whose tails extend into the visible region. An additional absorption peak is present near 1400 nm caused by OH^- absorptions in the fiber. This peak is due to the presence of residual water vapor in silica. It can be eliminated by taking special precautions in the production of the fiber [3]. The cumulated absorption spectrum is shown in Fig. 2.7.

Apart from attenuation, dispersion is another linear effect in signal propagation through a fiber. Dispersion refers to the phenomenon where different components of the signal travel at different group-velocities. As in this book solely single mode fibers are investigated, only a single mode is guided and no intermodal dispersion occurs. In fiber optics, however, the dependence of the group velocity on the wavelength plays an important role; therefore it is also called group velocity dispersion (GVD). In a single mode fiber, there are two main reasons for chromatic dispersion: material dispersion and waveguide dispersion. The origin of material dispersion is that the refractive index of silica is frequency-dependent. On the other hand, there is a component called waveguide dispersion. This component stems from the fact that the light is propagating partly in the core and partly in the cladding, and the power distribution between the core and the cladding is itself a function of the wavelength. Whereas material dispersion is more or less fixed, because silica is used as the material, waveguide dispersion can be tuned by changing the profile of the refractive index in a fiber. By deliberately changing this profile, different fiber types such as DSFs and DCFs can be produced [4].

The propagation constant $\beta(\omega)$, which is frequency-dependent, can be expanded in a Taylor series:

$$\beta(\omega) = \beta(\omega_0) + \frac{d\beta}{d\omega}\bigg|_{\omega_0}(\omega-\omega_0) + \frac{1}{2}\frac{d^2\beta}{d\omega^2}\bigg|_{\omega_0}(\omega-\omega_0)^2 + \frac{1}{6}\frac{d^3\beta}{d\omega^3}\bigg|_{\omega_0}(\omega-\omega_0)^3 + \ldots \quad (3.1\text{-}4)$$

Usually the derivatives of the propagation constant to the angular frequency are abbreviated by

$$\frac{d^n\beta}{d\omega^n}\bigg|_{\omega_0} = \beta_n. \quad (3.1\text{-}5)$$

The first element of the Taylor series expansion β_0 induces a frequency-independent phase response. The group velocity of the signal v_g is given by the first derivation of β [2]:

$$\left(v_g(\omega_0)\right)^{-1} = \frac{d\beta(\omega)}{d\omega}\bigg|_{\omega_0} = \beta_1. \quad (3.1\text{-}6)$$

This term can be eliminated in simulations by a moving coordinate system, which is assumed in the following. The Taylor series in Eq. 3.1-4 may be truncated after the second term for transmission systems covering only a narrow spectral width (e.g. single channel transmission of a 10 Gb/s transmission system). In today's state-of-the-art WDM systems with a large spectral bandwidth, however, it is mandatory to additionally consider the third derivative of the propagation constant. In future transmission systems with very high bit rates per channel also the inclusion of the fourth order dispersion term might be needed, especially if second- and third order dispersion are compensated [5].

The derivative of the group velocity in the spectrum is described by the dispersion parameter D and the dispersion slope parameter S [2]:

$$D = \frac{d}{d\lambda}\frac{1}{v_g} = -\frac{2\pi c}{\lambda^2}\beta_2 \qquad S = \frac{dD}{d\lambda} = \frac{(2\pi c)^2}{\lambda^3}\left(\frac{1}{\lambda}\beta_3 + \frac{1}{\pi c}\beta_2\right)$$
$$\beta_2 = \frac{d\beta_1}{d\omega} = \frac{d}{d\omega}\frac{1}{v_g} = -\frac{\lambda^2}{2\pi c}D \qquad \beta_3 = \frac{d\beta_2}{d\omega} = \frac{\lambda^3}{(2\pi c)^2}(\lambda S + 2D).$$

(3.1-7)

The key parameter describing the effect of GVD is the second derivative β_2 of the propagation constant with respect to the angular frequency. β_2 (ps^2/km) is called the GVD parameter. If β_2 is larger than 0, the chromatic dispersion is said to be normal. On the other hand, if $\beta_2 < 0$, the chromatic dispersion is called anomalous. Anomalous dispersion arises on a standard fiber operated in the C-band.

Accumulated dispersion is normally compensated by lumped elements of DCFs, which are deployed after every fiber span. In many DCFs the dispersion slope does not exactly match the slope of the transmission fiber, resulting in the imperfect compensation of GVD for frequencies lying far apart from the center frequency (e.g. the edge frequencies of a WDM signal). It is especially important to consider this effect for systems using both C- and L-bands. In these systems, an uncompensated distance of several kilometers may occur at the end of a long-haul transmission, causing severe system degradation due to uncompensated GVD. Newer DCF modules, however, show an improved dispersion slope matching today.

The imperfect dispersion compensation of edge channels may also be overcome by a channel-by-channel post compensation of the dispersion using tunable dispersion compensators, which are usually integrated today in the DSPs of coherent transmission systems. It is also possible to use other optical devices to compensate for the accumulated GVD. In recent years especially (chirped) fiber Bragg gratings have attracted interest because of their low price and low insertion loss, which is also independent of the amount of GVD to be compensated. Unfortunately group-delay ripple is inherent to such devices stemming from the production process. In the recent past several research papers have been published on how to overcome these distortions by using equalization techniques (e.g. [6–8]).

3.1.2 Amplified Spontaneous Emission Noise

In long-haul transmission systems optical amplifiers are employed to compensate for the fiber loss (compare also Sect. 2.5). These Erbium-doped fiber amplifiers are deployed every 80–120 km and increase the optical power launched into the amplifier by a certain gain G. The amplification principle of EDFAs is based on stimulated emission. A side effect of the amplification process is that spontaneous emission is added to the signal (compare also Fig. 2.7). In contrast to the stimulated photons, the photons generated by spontaneous emission have a random phase and polarization. Unfortunately, the spontaneous emission is also amplified along the Erbium-doped fiber, which leads to so called amplified spontaneous emission (ASE).

As a figure of merit for the degradation of an electrical signal by noise added during optical transmission the optical signal-to-noise ratio (OSNR) is used. For the definition of the OSNR it is assumed that an electrical signal is transmitted through an optical link and converted back at the receiver to the electrical domain. The OSNR is defined as [9]:

$$OSNR = \frac{\bar{I}^2}{\sigma^2} = \frac{I^2}{\sigma^2}, \quad (3.1\text{-}8)$$

where I is the expectation value of the photo current and σ^2 the variance of the current. The average photo current I of an ideal photo detector (with unity quantum efficiency) is given by [2]

$$I = \frac{P_S}{h \cdot f} e^-. \quad (3.1\text{-}9)$$

In Eq. 3.1-9 P_S is the average optical power, h is Planck's constant ($6.6256 \cdot 10^{-34}$ Js), f the optical frequency and e^- the elementary charge of an electron.

Through the quantization of the electrical charge shot-noise is introduced with [2]

$$\sigma^2 = 2e^- \cdot I \cdot B, \quad (3.1\text{-}10)$$

where B is the single-sided receiver bandwidth. By inserting Eqs. 3.1-9 and 3.1-10 into Eq. 3.1-8, the OSNR for shot-noise limited transmission can be calculated by

$$OSNR = \frac{P_S}{2h \cdot f \cdot B}. \quad (3.1\text{-}11)$$

When a signal without ASE noise is amplified by an EDFA, the OSNR at the output can be defined as [2]

$$OSNR = \frac{P_{out}}{P_{ASE}} \quad (3.1\text{-}12)$$

where P_{ASE} is the power of the ASE noise that is generated by the EDFA.

The noise power spectral density N_{ASE} (per polarization mode) of an EDFA can be expressed by [9]

$$N_{ASE} = n_{sp} \cdot h \cdot f_0 \cdot (G-1) \quad (3.1\text{-}13)$$

where n_{sp} is the spontaneous emission factor of the amplifier, f_0 is the reference frequency, and G is the amplifier gain. The spontaneous emission factor (or the population inversion factor) is given by [2]

$$n_{sp} = \frac{N_2}{(N_2 - N_1)} \quad (3.1\text{-}14)$$

where N_1 and N_2 are the atomic population densities of the ground and excited states, respectively. In Eq. 3.1-14 it is assumed that the emission- and absorption cross sections are identical and that the atomic population densities are independent of the frequency.

The optical power of the noise (for both polarizations) within a reference bandwidth B_0—usually chosen to be 0.1 nm—is given by [9]:

$$P_{ASE} = 2 \cdot N_{ASE} \cdot B_0. \quad (3.1\text{-}15)$$

With the help of Eq. 3.1-15 the OSNR after amplification can be expressed by:

$$OSNR = \frac{P_{out}}{2 n_{sp} \cdot h \cdot f_0 \cdot B_0 \cdot (G-1)}. \quad (3.1\text{-}16)$$

An amplifier can be characterized by a noise figure F. Assuming a signal containing only shot noise (no ASE) at the input and neglecting ASE-shot noise and ASE-ASE-beat noise at the output of the amplifier the noise figure can be expressed as [4, 9]

$$F = \frac{SNR_{in}}{SNR_{out}} = \frac{2 n_{sp}(G-1) + 1}{G} \quad (3.1\text{-}17)$$

where both SNR values are electrical signal-to-noise ratios, determined after converting the optical to an electrical signal with a photodiode. In the high gain region C-band (double-stage) amplifiers commonly have a noise figure of 4–6.5 dB, whereas L-band amplifiers have a higher noise figure of approximately 5–8 dB. This stems from the fact that in the L-band a higher pump power is needed, and the Erbium-doped fiber is much longer.

By inserting Eq. 3.1-17 into Eq. 3.1-16 the OSNR after amplification can be calculated by

$$OSNR = \frac{P_{out}}{(F \cdot G - 1) h \cdot f_0 \cdot B_0} = \frac{P_{in} \cdot G}{(F \cdot G - 1) h \cdot f_0 \cdot B_0}. \quad (3.1\text{-}18)$$

Equation 3.1-18 shows that the OSNR decreases with the amplifier gain. The higher an amplifier gain is required, the more ASE is added during the

3.1 Modeling of Fiber Optical Transmission Systems

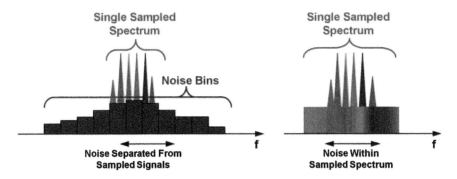

Fig. 3.1 Visualization of separate noise and sampled signal representations (*left*) and single sampled signal representation containing noise (*right*)

amplification process. Signals with a low optical power typically require a higher amplifier gain. Therefore it can be concluded that the lower the input power is, the higher the OSNR degradation of the signal will be.

In long-haul transmission systems consisting of a concatenation of several fiber spans, many EDFAs are cascaded to compensate for the losses. Under the assumption that all spans have identical loss, the ASE noise after N_spans with identical amplifiers can be written as:

$$P_{ASE} = N_{spans} \cdot P_{ASE,\text{single amplifier}}. \qquad (3.1\text{-}19)$$

In simulations, where noise is added numerically to the sampled signal, it can be modeled by an additive complex Gaussian noise process n.

$$A_{noise}(t) = A(t) + n \qquad (3.1\text{-}20)$$

The noise variance σ^2_{ASE} is related to the instantaneous power of the random process, which can be calculated from its noise power density S_{ASE} and its bandwidth B_0.

$$\sigma^2_{ASE} = P_{ASE} = S_{ASE} \cdot B_0 \approx \text{h} \cdot f_0 \cdot G \cdot F \cdot B_0 \qquad (3.1\text{-}21)$$

In Eq. 3.1-21 F is the noise figure of the amplifier, Planck's constant is denoted by h, and the gain is labeled G. G has been assumed to be large ($G \gg 1$). In addition, the gain function has been assumed flat, and the frequency dependence has been neglected. This is justified because the bandwidth of the amplifier is much smaller than the carrier frequency.

In numerical simulations ASE noise can be added to the sampled signal, or it may be treated separately (Fig. 3.1). In the former case a random realization of ASE noise with the given parameters is added to the sampled signal at each EDFA. This method allows the simulation of a most realistic system including the interaction of noise with the signal and also with fiber nonlinearities. However, the simulation of a very long bit sequence may be required to obtain reliable results on the bit error ratio. Usually Monte-Carlo (MC) simulations are used.

If noise is kept apart from the signal in a separate noise vector, signal and noise will not be combined before the end of a simulation when the bit error ratio is calculated (semi-) analytically (compare Sect. 3.1.8). This method has several advantages. First, a much shorter bit sequence is required for the (time consuming) numerical simulation of the signal. The bit sequence can be relatively short because only the influence of the linear and nonlinear fiber effects needs to be modeled. Additionally, the spectral width of the sampled signal can be significantly reduced because for signal and noise representations different spectral ranges can be employed. ASE-noise typically covers more than 5 THz, which is very inefficient to model on a sampled signal basis, if the signal uses only a fraction of this large bandwidth. Furthermore, noise can be assumed to be homogenously distributed over a relatively high bandwidth because it is essentially white noise, which is filtered along the transmission. This is why coarse noise bins for the noise representation can be used, which however cover in total a large frequency range (compare Fig. 3.1, left). In these bins noise can be represented by the average spectral density (in two polarizations).

3.1.3 Fiber Birefringence

The effects of fiber birefringence have been neglected deliberately so far. This simplification is only valid for fibers with a low birefringence and transmission systems operating with low bit rates.

In single mode fibers generally two fundamental modes, which are orthogonal to each other, are guided. In the degenerate case (assumed so far in this chapter), both modes have the same group velocities, which means that the propagation constants β for both modes are identical. In reality, though, the propagation constants of the axes differ because of e.g. an asymmetrical fiber core or mechanical stress applied to the fiber [2]. If an input pulse excites both polarization components, it becomes broader after transmission as the two components disperse along the fiber because of their different group velocities. This effect is called polarization mode dispersion (PMD). In polarization maintaining fibers the birefringence is constant. In conventional fibers, however, birefringence varies along the fiber in a random fashion (compare also Fig. 3.3). It is intuitively clear that the polarization state of light propagating in such fibers with randomly varying birefringence will generally be elliptical and will change randomly along the fiber during propagation. The polarization state will also vary for different spectral components of the pulse.

The difference of the propagation constants for both orthogonal modes can be written as [2]:

$$\Delta\beta_1 = |\beta_{1,x} - \beta_{1,y}|. \qquad (3.1\text{-}22)$$

This leads to the propagation equation for a linear, birefringent fiber

3.1 Modeling of Fiber Optical Transmission Systems

$$\frac{\partial}{\partial z}\begin{pmatrix} A_x(z,t) \\ A_y(z,t) \end{pmatrix} + \left(\frac{\alpha}{2} - \frac{1}{2}\begin{pmatrix} \Delta\beta_1 & 0 \\ 0 & -\Delta\beta_1 \end{pmatrix}\frac{\partial}{\partial t} + \frac{j}{2}\beta_2\frac{\partial^2}{\partial t^2} - \frac{1}{6}\beta_3\frac{\partial^3}{\partial t^3}\right)\begin{pmatrix} A_x(z,t) \\ A_y(z,t) \end{pmatrix} = 0.$$

(3.1-23)

How a transmission system with randomly varying birefringence can be modeled is shown in Sect. 3.1.7.

3.1.4 Nonlinear Fiber Effects

To this point, the nonlinear fiber effects have been omitted. Consequently, Eqs. 3.1-2 and 3.1-23 may only be used for very low input powers, where the nonlinear effects are negligible. In the following, also the nonlinear fiber effects are included. The cause of fiber nonlinearity is that the response of any dielectric (in this case: the fiber) to light becomes nonlinear for intense electromagnetic fields. These nonlinear effects have an increasing importance in optical communications because of the considerable high aggregate input powers of today's transmission systems with a large number of WDM channels and very long transmission distances. Even though silica is intrinsically not a highly nonlinear material, the waveguide geometry that confines light to a small section leads to a high power density [2].

In a fiber there exist elastic and inelastic scattering processes. The elastic scattering processes stem from the Kerr effect. This effect describes the dependence of the refractive index on the instantaneous power of the transmission fiber (Eq. 3.1-25). The Kerr effect is energy conserving regarding the optical energy. As a result, it distorts the phase of an optical signal.

If only a single channel is transmitted, the Kerr effect leads to self-phase modulation (SPM). The time-dependent power of the optical signal (i.e. the rising and falling edges) causes a phase shift. This phase shift is transferred to an amplitude shift by group-velocity dispersion (GVD) [1]. In the anomalous dispersion regime, the phase shifts of SPM and GVD have opposite signs. These properties are used in soliton systems, where the effect of SPM exactly compensates the signal degradation due to GVD.

In WDM systems another effect becomes important. This effect is called crossphase modulation (XPM). Through XPM a WDM channel with (time-) varying intensity induces a phase shift on other WDM channels because the refractive index of the fiber is changed by the instantaneous powers of the other channels [1]. Due to the usually high number of WDM channels and the modulated nature of the channels, the degradation due to XPM can only be described by a stochastic process. In intensity modulated transmission systems XPM leads to noise-like crosstalk between the WDM channels. In phase modulated systems phase noise is induced by XPM, which is especially strong, if NRZ-OOK neighboring channels are present in the system. Recently also polarization crosstalk induced by XPM (also referred to as XPolM or PolXPM) has come into the focus. This effect leads to crosstalk between the different components of a signal transmitted on orthogonal polarization states (for example using the modulation format polmux QPSK). Additional information can be found in e.g. [10, 11] or [12].

Another elastic effect is four-wave mixing (FWM). The term four-wave mixing comes from its property that three propagating waves generate a fourth one. This is because of the nonlinear phase shift, which is determined by the square of the complex field amplitudes. By squaring the aggregate field amplitude, mixing products are generated. This effect is most detrimental in WDM systems with equidistant frequency spacing and a low local dispersion. In this case, the mixing products fall into the signal channels and lead to crosstalk. For the development of FWM crosstalk phase matching is crucial. This means that the channels, which are involved in the FWM process, need to propagate for a certain distance with the same phase relation. In the case of DSF, FWM leads to significant signal degradation because of very low local dispersion and nearly perfect phase matching. For higher dispersion fibers such as NZDSFs and especially SSMFs, the effect of FWM becomes less strong because a high walk-off between the different channels exists.

There are also inelastic scattering processes. Stimulated Raman scattering (SRS) and stimulated Brillouin scattering (SBS) belong to this group. In the case of SBS, the scattering of the pump channel is caused by acoustic waves, which lead to a contra-directional frequency-shifted Stokes wave. SRS, on the other hand, is caused by optical phonons. A Stokes wave, which is also shifted in frequency, will be amplified or can even be generated out of the noise. In contrast to SBS, the Stokes wave caused by SRS travels in the same direction as the signal. The SBS process can be effectively eliminated by modulation of the lasers with a low frequency (dither).

In comparison to the elastic scattering processes, in inelastic scattering processes the number of photons is conserved instead of the optical energy. This stems from the fact that energy is exchanged with acoustical or optical molecule oscillations (phonons).

For higher bit rate systems intrachannel effects become dominant (compare e.g. [13, 14]). Intra-channel nonlinear effects already occur in single channel transmission. The reason for intra-channel FWM (I-FWM) is that optical pulses are broadened by GVD in the fiber and overlap each other. When these pulses, which have different overlapping optical frequency components, are propagating through a fiber, the different spectral components interact and I-FWM occurs. Because the difference in the optical frequencies is smaller than the spectral width, FWM efficiency becomes sufficiently high for I-FWM light to appear. I-FWM light can be seen at the temporal position '0' (of an OOK signal) as ghost-pulses in an eye diagram. It can be regarded as channel crosstalk which is one reason for the power penalty. Intrachannel effects have to be considered, if the shortening of the pulses (i.e. in high bit rate transmission systems $\geq$ 40 Gb/s) makes the dispersion length l_D in the fiber much shorter than the nonlinear length l_{NL} [1].

$$l_D = \frac{T_0^2}{|\beta_2|}, \quad l_{NL} = \frac{1}{\gamma \cdot P_0}. \tag{3.1-24}$$

3.1 Modeling of Fiber Optical Transmission Systems

In Eq. 3.1-24 T_0 defines the input pulse width, P_0 is the input power and γ the nonlinearity constant (compare Eq. 3.1-26). The intensity pattern therefore changes rapidly along the fiber. Note that this scheme is in some sense opposite to soliton or dispersion managed soliton transmission. The concept of combating intrachannel crosstalk is spreading the pulses as far as possible and as quickly as possible in the time domain, creating a rapidly varying intensity pattern.

Another detrimental effect is the effect of intrachannel XPM (I-XPM). I-XPM results in timing jitter of the marks. In [15] it has been found out that I-XPM may become dominant for 40 Gb/s (RZ-OOK) systems employing SSMF, if no pre-compensation of dispersion is employed.

The dependence of the refractive index on the optical power is described by the Kerr effect shown in the following equation [2]:

$$n(t) = n_{linear} + n_2 |A(t)|^2. \qquad (3.1\text{-}25)$$

The nonlinear refractive index n_2 in Eq. 3.1-25 is related to the nonlinear coefficient γ of the fiber by [2]:

$$\gamma = \frac{n_2 \cdot \omega_0}{c_0 \cdot A_{eff}}. \qquad (3.1\text{-}26)$$

In this equation c_0 ($2.99792 \cdot 10^8$ m/s) stands for the speed of light in vacuum, and A_{eff} is the effective area of the fiber core. The nonlinear coefficient varies between 1.29 (W·km)$^{-1}$ for an NZDSF and approximately 5.0 (W·km)$^{-1}$ for a DCF because of the much smaller core diameter of the DCF.

An optical fiber with large birefringence can be modeled by the following equation including the nonlinear elastic effects, dispersion and attenuation [1]:

$$\frac{\partial}{\partial z}\begin{pmatrix} A_x \\ A_y \end{pmatrix} + \left\{ \frac{\alpha}{2} - \frac{1}{2}\begin{pmatrix} \Delta\beta_1 & 0 \\ 0 & -\Delta\beta_1 \end{pmatrix} \frac{\partial}{\partial t} + \frac{j}{2}\beta_2 \frac{\partial^2}{\partial t^2} - \frac{1}{6}\beta_3 \frac{\partial^3}{\partial t^3} \right\} \begin{pmatrix} A_x \\ A_y \end{pmatrix}$$
$$= j\gamma \begin{pmatrix} |A_x|^2 + \frac{2}{3}|A_y|^2 & 0 \\ 0 & |A_y|^2 + \frac{2}{3}|A_x|^2 \end{pmatrix} \begin{pmatrix} A_x \\ A_y \end{pmatrix} \qquad (3.1\text{-}27)$$

In this context large birefringence means that the beat length l_B is much smaller than the fiber length l, where the beat length is defined as:

$$l_B = \frac{2\pi}{|\beta_x - \beta_y|}. \qquad (3.1\text{-}28)$$

For typical fibers the beat length l_B is around 1 m. If, however, fibers with very low birefringence are assumed, additional crosstalk terms need to be added to Eq. 3.1-27, which are averaging to zero in the case of high birefringence. This leads to the following coupled-mode equation for low birefringence fibers:

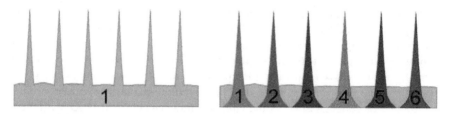

Fig. 3.2 Visualization of the total field approach, where all channels are included in one sampled signal (*left*) and separated channel approach, where for each channel a sampled signal vector is generated (*right*)

$$\frac{\partial}{\partial z}\begin{pmatrix}A_x\\A_y\end{pmatrix}+\left\{\frac{\alpha}{2}-\frac{1}{2}\begin{pmatrix}\Delta\beta_1 & 0\\0 & -\Delta\beta_1\end{pmatrix}\frac{\partial}{\partial t}+\frac{j}{2}\beta_2\frac{\partial^2}{\partial t^2}-\frac{1}{6}\beta_3\frac{\partial^3}{\partial t^3}\right\}\begin{pmatrix}A_x\\A_y\end{pmatrix}$$
$$=j\gamma\begin{pmatrix}|A_x|^2+\frac{2}{3}|A_y|^2 & \frac{1}{3}A_x^*A_y\exp(-2j\Delta\beta z)\\\frac{1}{3}A_y^*A_x\exp(-2j\Delta\beta z) & |A_y|^2+\frac{2}{3}|A_x|^2\end{pmatrix}\begin{pmatrix}A_x\\A_y\end{pmatrix} \qquad (3.1\text{-}29)$$

For ultra short pulses (<5 ps), which may occur, if a WDM transmission system with a high channel count and very high bit rate is simulated, additional terms have to be included taking into account both the electronic and vibrational (Raman) contributions [1]:

$$\frac{\partial}{\partial z}A_i+\frac{\alpha}{2}A_i\mp\frac{1}{2}\Delta\beta_1\frac{\partial A_i}{\partial t}+\frac{j}{2}\beta_2\frac{\partial^2 A_i}{\partial t^2}-\frac{1}{6}\beta_3\frac{\partial^3 A_i}{\partial t^3}$$
$$=j\gamma\left\{|A_i|^2A_i+\frac{j}{\omega_0}\frac{\partial}{\partial t}\left(|A_i|^2A_i\right)+\frac{1}{3}A_i^*A_j^2+\frac{2}{3}|A_j|^2A_i-(\tau_{R1}+\tau_{R2})A_i\frac{\partial}{\partial t}|A_i|^2-\tau_{R2}A_i\frac{\partial}{\partial t}|A_j|^2\right\}$$
$$(3.1\text{-}30)$$

with the parallel and orthogonal Raman time constants τ_{R1} and τ_{R2} and A_i respectively A_j being the wave envelopes of the two orthogonal linearly polarized modes. Equation 3.1-30 can be used to model simulation systems with a pulse width as low as 10 fs [1].

3.1.5 Coupled Nonlinear Schrödinger Equation

Two possible ways of modeling the NLSE exist. Either the so-called total field (TF) model can be applied, or the separated channels (SC) model can be used. Up to this point the equations have been given for the TF model only. In this model the electrical fields of all channels are added together and are simulated simultaneously in a single sampled vector (Fig. 3.2, left). Especially for transmission systems with a high spectral separation between the different channels, however, the SC model is more suited because unused bands can be left away in the simulation (Fig. 3.2, right). In SC mode the (nonlinear) interaction between the

3.1 Modeling of Fiber Optical Transmission Systems

individual channels is obtained by coupling terms. This also allows switching on and off the different nonlinear degradation effects separately enabling a more detailed analysis of the individual contributions.

In the SC case the total electric field amplitude is given by the sum of the separate electric fields of the individual WDM channels.

$$A = \sum_{n=1}^{N} A_n \qquad (3.1\text{-}31)$$

The total number of channels is denoted by N, and the amplitudes (of the envelopes) of the different channels are referred to as A_n. The indices written in subscript refer to the channel numbers.

The evolution of the amplitudes of the individual optical signals along the fiber is modeled by the coupled NLSE [1]. To increase clarity the polarization effects are not taken into account in Eq. 3.1-32, and only one polarization axis is assumed.

$$\frac{\partial A_n}{\partial z} + \underbrace{\frac{\alpha}{2} A_n}_{\text{Attenuation}} + \underbrace{\left(\beta_{1,n} - \beta_{1,ref}\right) \frac{\partial A_n}{\partial t}}_{\text{Time delay}} + \underbrace{\frac{j}{2} \beta_{2,n} \frac{\partial^2 A_n}{\partial t^2}}_{\text{Dispersion}} - \underbrace{\frac{1}{6} \beta_{3,n} \frac{\partial^3 A_n}{\partial t^3}}_{\text{Dispersion slope}}$$

$$= + j\gamma A_n \left\{ \underbrace{|A_n|^2}_{\text{SPM}} + 2 \underbrace{\sum_{i=1, i \neq n}^{N} |A_i|^2}_{\text{XPM}} \right\} + j\gamma \underbrace{\sum_{n=i+j-k; i,j \neq k} A_i A_j A_k^* \exp(-j\Delta k z)}_{\text{FWM}}$$

$$\underbrace{\left(-\sum_{i=1}^{n-1} \frac{\omega_n}{\omega_i} \frac{g_R}{2K \cdot A_{eff}} |A_i|^2 + \sum_{i=n+1}^{N} \frac{g_R}{2K \cdot A_{eff}} |A_i|^2 \right) A_n}_{\text{SRS}}$$

$$(3.1\text{-}32)$$

The phase constant Δk is given by:

$$\begin{aligned}\Delta k &= k_i + k_j - k_k - k_n \\ &= -\beta_2(\omega_i - \omega_k)(\omega_j - \omega_k) - \beta_3(\omega_i - \omega_k)(\omega_j - \omega_k)\left(\left(\omega_i + \omega_j\right)/2 - \omega_0\right)\end{aligned}$$

$$(3.1\text{-}33)$$

On the left hand side of Eq. 3.1-32 the linear fiber effects, which are attenuation and dispersion, are specified. Furthermore, a time delay is included, which stems from the different group velocities of the separate channels. $\beta_{1,ref}$ refers to the propagation constant of the channel, which is selected to represent the retarded time frame. The nonlinear terms are defined on the right hand side of the equation. The effects of XPM and SRS lead to coupling between the different channels. In this context g_R is the Raman gain factor and K is the polarization factor. For co-polarization of the WDM channels K equals 1, and for completely random polarization of the channels K equals 2. In the SC model, the FWM effect generates new channels.

For the efficient simulation of FWM, only the crosstalk of the FWM mixing products into existing channels ought to be simulated. If the power transfer into neighboring frequencies should also be simulated, which is usually very small, additional dummy channels may be added, which do not carry optical signals.

3.1.6 Split-Step Fourier Method

As already mentioned above, the NLSE can be solved analytically in special cases only. This is why numerical methods are used to find a solution. In this book, the split-step Fourier method (SSFM) has been employed. The SSFM allows one to solve the nonlinear Schrödinger equation efficiently and with a high accuracy. Finding a solution to the NLSE is not a trivial problem. Several different methods for solving the NLSE are well established, and it remains an interesting topic of recent research activities (compare e.g. [16–21]). In the past, several attempts have been made to speed up SSFM-based simulations e.g. by introducing IIR filter structures [19] that approximate the linear operator of the nonlinear Schrödinger equation or by a split-step wavelet collocation method [17]. Most commercial simulation tools, however, are still based on the standard SSFM because it features the highest reliability.

Recently parallelization techniques have become popular due to the availability of cheap parallel processors in the form of graphics processing units. More information on this topic can be found in Chap. 4 of this book.

The split-step approach divides the NLSE into a linear part $\hat{L}$ and a nonlinear part $\hat{N}$. Both portions are solved independently. This is a good approximation for small discretization steps Δz, where linear and nonlinear parts do not interact in first approximation.

The linear part is—according to total-field Eq. 3.1-27—given by:

$$\hat{L} = \frac{1}{2} \begin{pmatrix} \Delta\beta_1 & 0 \\ 0 & -\Delta\beta_1 \end{pmatrix} \frac{\partial}{\partial t} - \frac{j}{2}\beta_2 \frac{\partial^2}{\partial t^2} + \frac{1}{6}\beta_3 \frac{\partial^3}{\partial t^3} \qquad (3.1\text{-}34)$$

and the nonlinear part including the attenuation of the signal is defined as:

$$\hat{N} = -\frac{\alpha}{2} + j\gamma \begin{pmatrix} |A_x|^2 + \frac{2}{3}|A_y|^2 & 0 \\ 0 & |A_y|^2 + \frac{2}{3}|A_x|^2 \end{pmatrix}. \qquad (3.1\text{-}35)$$

As a simplified notation, the differential equation can be noted as:

$$\frac{\partial}{\partial z} \begin{pmatrix} A_x \\ A_y \end{pmatrix} = (\hat{L} + \hat{N}) \begin{pmatrix} A_x \\ A_y \end{pmatrix} \qquad (3.1\text{-}36)$$

where the attenuation can either be computed in the linear or the nonlinear part. The linear part $\hat{L}$ is solved in the frequency domain, and the nonlinear part $\hat{N}$ is solved in the time domain. The transformation between time and frequency domains and vice versa is obtained by fast Fourier transforms (FFTs) or IFFTs, respectively. As both parts are treated separately only small step-sizes are admissible. From the physical point of view, though, there is an interaction of both portions during a split-step. The error, however, can be kept below a desired level, if the step-size is chosen accordingly.

For the efficiency of the SSF method, the right choice of the step-size is crucial. If the step-size is chosen too large, the results will be inaccurate due to insufficient consideration of the interaction of the linear and nonlinear operators. On the other hand, if the step-size is chosen too small, the computational efficiency is decreased. Typically, several criteria are used to limit the maximum step-size, and always the strictest one (leading to the smallest step-size) is used. These criteria include a maximum nonlinear phase shift between two split-steps, a maximum amount of artificial FWM and a maximum walk-off between two WDM channels.

The nonlinear phase shift is defined as follows:

$$\varphi_{NL,max} = \gamma |A_{max}|^2 \Delta z. \quad (3.1\text{-}37)$$

For a high accuracy typically a maximum nonlinear phase shift between two split-steps of less than 0.1 mrad is required. Equation 3.1-37 is dependent on the maximal amplitude of the signal. Due to the attenuation of the fiber the amplitude is decreasing exponentially. This enables one to increase the step-sizes while approaching the end of the fiber.

Another important aspect, which needs to be considered, is the influence of the SSF method on the FWM effect. If the split-steps have a constant size, FWM is overestimated significantly. This can be reduced considerably by small steps or alternating step-sizes [22]. By adapting the step size to the attenuation profile of the fiber, a variation in the split-step length is automatically gained. The reason for the overestimation of the FWM effect is the operator splitting. As the nonlinear operator is applied, the effects of dispersion are neglected. This leads to a phase matching, as it would occur on a DSF fiber, and there will be a high efficiency of the FWM. For small step-sizes this effect can be overcome. If the steps have the same length, the FWM mixing products will add-up coherently. In the literature, an algorithm has been published which allows one to calculate the split-steps for a maximal given artificial FWM threshold [23]. This criterion is used to bound the step-size in addition to the nonlinear phase shift. Usually at least 30 dB artificial FWM suppression is required for a high accuracy of the simulations.

In the simulation tool *PHOTOSS*, asymmetrical operator splitting is implemented. This means that first the linear and then the nonlinear operator is applied, always in that order. On the other hand, symmetrical operator splitting could be utilized [1].

$$A(z+\Delta z, T) = \exp\left(\frac{1}{2}\hat{L}\Delta z\right) \exp\left(\int_{z}^{z+\Delta z} \hat{N}(\tilde{z})d\tilde{z}\right) \exp\left(\frac{1}{2}\hat{L}\Delta z\right) \cdot A(z,T) \quad (3.1\text{-}38)$$

For the transformation of the signal between time and frequency domains a fast Fourier transform (FFT) is employed as mentioned above. The total computational time of the SSFM is dominated by the computational time of the FFTs (and IFFTs) and can be significantly enhanced by a fast parallel implementation of the FFT (shown in Chap. 4).

3.1.7 Modeling of Polarization Mode Dispersion

As already presented in Sect. 3.1.3, a single mode fiber is generally birefringent with different propagation constants $\beta_x(\omega) \pm \beta_y(\omega)$ of the two orthogonal axes. The differences stem from disruptions of the fiber symmetry during the production process. Furthermore, mechanical and thermal stress, vibrations, as well as bending or torsion, lead to statistical coupling of the two propagation modes. Generally the birefringence is varying randomly along the fiber. This leads to so called polarization mode dispersion (PMD).

Mathematically, a 2-D expression of polarization using the electric field vector is known as the Jones vector representation [24], where E_x, E_y, φ_x and φ_y are all real numbers.

$$\vec{s} = \begin{pmatrix} E_x e^{j\varphi_x} \\ E_y e^{j\varphi_y} \end{pmatrix} \frac{1}{\left|\sqrt{E_x^2 + E_y^2}\right|} = \begin{pmatrix} s_x \\ s_y \end{pmatrix} \tag{3.1-39}$$

The Jones vector describes the state of polarization of the light. Another way of describing the polarization of light is the Stokes vector representation [25]. It consists of four real parameters, where I stands for the optical power:

$$\begin{array}{ll} S_0 = I & S_1 = I_x - I_y \\ S_2 = I_{45°} - I_{-45°} & S_3 = I_{RHC} - I_{LHC}. \end{array} \tag{3.1-40}$$

The abbreviations RHC and LHC stand for right hand circular and left hand circular, respectively. The normalized Stokes vector is given by:

$$\hat{s} = \{S_1, S_2, S_3\}/S_o = \{s_1, s_2, s_3\}. \tag{3.1-41}$$

With the help of the Pauli spin matrices it is possible to obtain a Stokes vector from the Jones vector [26]

$$s_i = \vec{s}^* \cdot \sigma_i \cdot \vec{s} \tag{3.1-42}$$

The Pauli matrices are defined as follows [27]:

$$\sigma_1 = \begin{bmatrix} 0 & 1 \\ 1 & 0 \end{bmatrix}; \quad \sigma_2 = \begin{bmatrix} 0 & -j \\ j & 0 \end{bmatrix}; \quad \sigma_3 = \begin{bmatrix} 1 & 0 \\ 0 & -1 \end{bmatrix}. \tag{3.1-43}$$

In the so called "wave plate model" the fiber is modeled as a cascade of birefringent elements, which are rotated by random angles (Fig. 3.3). Apart from the angle, the wave plate length and phase of the mode coupling are also random variables. The mode coupling is described by the following transformation:

3.1 Modeling of Fiber Optical Transmission Systems

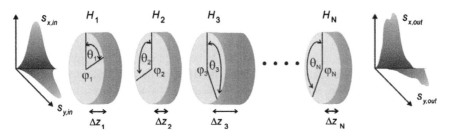

Fig. 3.3 Visualization of the wave plate model

$$\vec{s}_{out} = C \cdot D \cdot \vec{s}_{in} \quad (3.1\text{-}44)$$

with

$$C = \begin{bmatrix} \exp(-j\frac{\varphi}{2}) & 0 \\ 0 & \exp(j\frac{\varphi}{2}) \end{bmatrix} \quad (3.1\text{-}45)$$

and

$$D = \begin{bmatrix} \cos\Theta & \sin\Theta \\ -\sin\Theta & \cos\Theta \end{bmatrix}. \quad (3.1\text{-}46)$$

The mode coupling between the different elements is determined by the two angles Θ and φ. The matrices for the transformation of the polarization state are referred to as Jones matrices.

Fig. 3.3 depicts schematically the setup of the wave plate model. A single wave plate is assumed to have a birefringence $\Delta\beta_1 = \beta_{1,x} - \beta_{1,y}$, which is independent from the position. For wave plates with random lengths, the position of the mode coupling is obtained from an independent random process.

The Jones matrix for a single wave plate of length Δz_i, is given by

$$H(\omega) = D^{-1} \cdot B(\omega) \cdot C \cdot D \quad (3.1\text{-}47)$$

and

$$B(\omega) = \begin{bmatrix} \exp(-\frac{1}{2}\Delta\beta_1\omega\Delta z_i) & 0 \\ 0 & \exp(\frac{1}{2}\Delta\beta_1\omega\Delta z_i) \end{bmatrix} \quad (3.1\text{-}48)$$

According to the wave plate model, the fiber consists of a limited number of wave plates (e.g. 100), and the Jones matrix of the entire fiber $H_{tot}(\omega)$ is given by the product of the individual Jones matrices. The output signal $A_{out}(\omega)$ can be obtained in the frequency domain by a multiplication of the Jones matrix with the input signal:

$$\vec{s}_{out}(\omega) = \prod_{i=1}^{N} H_i(\omega) \cdot \vec{s}_{in} = H_{tot}(\omega) \cdot \vec{s}_{in}. \quad (3.1\text{-}49)$$

3.1.8 Calculation of the Bit Error Ratio

If the signal quality after transmission has to be evaluated, the bit error ratio (BER) is a commonly used metric. To calculate the BER the detected bit sequence can be compared to the bit sequence originally sent by the transmitter. For a reliable estimation of the BER at least 100 bit errors should be counted [28]. In modern long-haul transmission systems operating at a BER level below 10^{-12} this leads to more than 10^{14} bits that have to be simulated. This is not acceptable in many cases due to the excessive use of computational time needed to conduct these calculations. Therefore this approach, which is also known as the "Monte Carlo method", may only be employed in systems in which the expected BER is much higher, so that the number of bits to calculate is decreased significantly. This is why in many simulations (and also laboratory experiments) the so called pre-FEC BER is measured. Typically a BER of 10^{-3} is desired before FEC because it can be corrected by the standardized FEC to an error free level (BER $< 10^{-12}$). Still at least 10^5 bits must be simulated for this case.

Another way of calculating the BER of a system is the so-called "tail extrapolation" described in [29]. For this extrapolation it is necessary that the values for the marks and spaces are Gaussian distributed which is valid to a first approximation in many transmission systems [2]. Based on this assumption, the threshold that distinguishes the marks from the spaces can be altered, causing deliberately higher bit error ratio, which are connected to the according threshold. By sweeping the threshold up to the maximum and minimum values of the signal, the bit error ratio increases so much that it is possible to reduce the number of simulated bits to an acceptable amount. Since there is a functional relation between the threshold and the BER, the optimal threshold and the according BER can be estimated by extrapolating the results for higher and lower thresholds.

Another method for calculating the BER is a semi-analytical approach [30]. It is based on theoretical considerations about the noise induced into the system. Therefore the variances of the different noise contributions have to be calculated. The different noise contributions are listed in the following (assuming a PIN diode based receiver).

Due to the quantized nature of the light, shot noise arises in the electrical current. The shot noise shows a Poisson distribution (Eq. 3.1-10).

Furthermore, thermal noise is generated at the receiver. The variance of the thermal noise is given by [2]:

$$\sigma^2_{thermal} = \frac{4 k_B T}{R_L} \Delta f_{el} \qquad (3.1\text{-}50)$$

with the Bolzmann constant k_B (1,38065 · 10^{-23} J/K), the ambient temperature T and the load resistor of the photodetector R_L. Equation 3.1-50 is independent from the current i, therefore the variances of the marks and spaces are identical.

As previously mentioned EDFAs inherently generate ASE noise. When ASE noise is detected by the receiver, it is converted into shot noise [31].

3.1 Modeling of Fiber Optical Transmission Systems

$$\sigma^2_{ASE-shot} = 2 \cdot e^- \left(R \cdot S_{ASE} \cdot \Delta f_{opt} \right) \Delta f_{el} \tag{3.1-51}$$

The ASE-shot noise in Eq. 3.1-51 depends on the elementary charge e⁻, the responsitivity of the photodiode R (also referred to as the conversion efficiency), the noise power density S_{ASE} and the optical and electrical filter bandwidths of the optical demux filter and the photodiode's low pass characteristic, respectively. The ASE-shot noise term in Eq. 3.1-51 can be ignored in most cases since the ASE-ASE beat noise term (Eq. 3.1-52) is much larger.

Because the optical amplitude of the signal is squared at the photodiode—from the mathematical point of view—additional beat-noise is generated. This results from different spectral components of the noise. The beating term of the signal is small—compared to the beat noise of ASE terms—because of the much narrower spectral width of the signal. ASE-ASE beat noise is determined by:

$$\sigma^2_{ASE-ASE} = 2 \cdot (R \cdot S_{ASE})^2 \Delta f_{opt} \cdot \Delta f_{el}. \tag{3.1-52}$$

Also, ASE-ASE beat noise does not show a dependence on the signal amplitude.

Finally, beat noise between the signal and ASE noise is generated. The variance of ASE-channel noise is defined by:

$$\sigma^2_{ASE-channel} = 4 \cdot i \frac{R \cdot S_{ASE}}{2} \Delta f_{el}. \tag{3.1-53}$$

As in the case of ASE-ASE noise, ASE-channel noise is also exclusively generated by co-polarized signals. This means that only the signal parts, which have the same polarization, can lead to beat noise. A factor of 0.5 is introduced to take this effect into account.

Because the noise effects are uncorrelated, the variances can be added to yield the overall noise variance:

$$\sigma^2(i) = \sigma^2_{shot}(i) + \sigma^2_{ASE-shot} + \sigma^2_{ASE-ASE} + \sigma^2_{ASE-channel}(i) + \sigma^2_{thermal}. \tag{3.1-54}$$

In transmission system employing intensity modulation and direct detection ASE-channel beat noise is dominant. The other noise terms can usually be neglected.

If the noise is assumed to be Gaussian distributed and zeros and ones are equally probable, the BER can be calculated from the following equation:

$$BER = \frac{1}{2} \left[erfc\left(\frac{i_1 - i_{threshold}}{\sigma_1 \cdot \sqrt{2}} \right) + erfc\left(\frac{i_{threshold} - i_0}{\sigma_0 \cdot \sqrt{2}} \right) \right] \tag{3.1-55}$$

In Eq. 3.1-55 i_1 and i_0 are the photo currents of the marks and spaces, $i_{threshold}$ is the decision threshold distinguishing between the two signal levels, σ_1 and σ_0 are the standard deviations due to noise, and erfc is the error function complement.

If the decision threshold is chosen optimally, the BER can be calculated from:

$$BER = \frac{1}{2} erfc\left(\frac{Q}{\sqrt{2}} \right) \tag{3.1-56}$$

with

$$Q = \frac{i_1 - i_0}{\sigma_1 + \sigma_0} \qquad (3.1\text{-}57)$$

The semi-analytical approach outlined above is especially suited to the separation of noise and signal vectors as described in Sect. 3.1.2.

For balanced detection receivers a different approach has to be adopted to calculate the BER [32]. This time the exact probability density function (PDF) of the detection noise is needed rather than the Gaussian approximation to this PDF. The reason for this important simulation aspect is that the tails of the exact PDFs differ significantly from the tails of the Gaussian distribution. For single ended detection as used for OOK this difference in the PDFs cancels to a high degree of accuracy. This beneficial cancellation is not found in balanced receivers, though [32]. This is owing to the fact that in single-ended receivers the single, noisy signal is compared to a deterministic (non-noisy) threshold, whereas in a balanced receiver two noisy signals are compared to each other. Therefore the BER should be computed using the exact PDFs. For this purpose there exist several methods which are based on the so-called Karhunen-Loève (KL) series expansion model [33, 34]. The main idea is to find a set of orthonormal basis functions that make the expanded noise components statistically independent. By this method an accurate BER can be calculated taking into account ASE-noise, pulse shaping, optical and electrical filtering and interferometer phase error.

For today's state-of-the-art transmission systems with coherent detection and digital signal processing, yet another approach may be adopted. It is also based on a separate representation of noise and signal components. The idea is to combine analytical and numerical noise modeling. During the (time consuming) simulation of the fiber separate noise and sampled signal vectors are used. In front of the receiver the analytical noise representation is converted into a random noise realization and added to the sampled signal. Different noise realizations can be added in a MC-like fashion consecutively to the same sampled signal (which has been simulated only once for the transmission link). This enables simulating the fiber once with a relatively small number of bits required for the assessment of the signal degradation due to the linear and nonlinear effects along the transmission line, yet allows accurately modeling the coherent receiver and the DSP with a high number of bits (and different noise realizations).

3.2 The Fiber Optical Transmission Simulator PHOTOSS

For the numerical simulation of optical transmission systems, in this book solely the simulation software *PHOTOSS* has been used, which has been developed for several years at the Chair for High Frequency Technology of TU Dortmund, Germany [35]. With this simulation tool all the components contained in a fiber

3.2 The Fiber Optical Transmission Simulator PHOTOSS

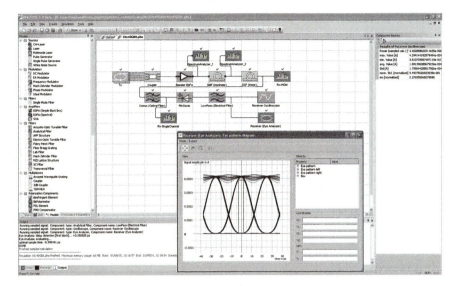

Fig. 3.4 Graphical user interface of the simulation tool *PHOTOSS*. The component library is depicted on the left hand side. The eye diagram depicts the received signal of the center WDM channel

optical transmission system can be modeled, starting from the transmitter up to the receiver. A graphical user interface (GUI) allows a comfortable and easy-to-use operation (compare Fig. 3.4). *PHOTOSS* is commercially distributed and also used by many project partners in research projects and the industry around the globe. Simulation tools for modelling the physical layer of fiber optical transmission systems are also distributed by e.g. [36–38].

A large component library is part of *PHOTOSS*. Pre-defined components can be dropped onto the simulation grid, and the parameters of the various components can be modified easily.

Also a parameter variation tool is part of *PHOTOSS*. The parameter variation allows one to vary certain system parameters during a simulation. This enables to investigate the influences of these parameters on the system performance. *PHOTOSS* performs separate simulations for all given parameters. After the simulation is finished, the results can be plotted against the varied parameters.

Another tool included in *PHOTOSS* is the so-called path analysis. With it the signal quality along a transmission path can be monitored, which may include various components in a network. The evaluated signal parameters may be—for example— the eye opening, the BER, the Q-factor or the OSNR. This function facilitates finding the optimum lengths of a DCF by determining the optimal residual dispersion, or it allows displaying the signal evolution along the fiber transmission line.

To simplify the setup of the optical transmission system, networks and iterators may be used. Networks allow grouping certain components. For example, all components belonging to a transmitter or receiver may be put into a network. Iterators can be used to repeat certain common operations. They may be used to

model a transmitter array consisting of identical lasers with the only varying parameter being the wavelength. Another important application for iterators is to provide a user-friendly way of modeling cascaded transmission spans.

Recently the functionality of *PHOTOSS* has been extended by integrating a scripting language (called PScript), which allows the user to embed simulations into an algorithmic context. In this way complex parameter variations or enhanced MC simulations (such as stratified-sampling) can be used easily.

Furthermore, a significant reduction of the computational time needed for the simulation of the fiber has been achieved by utilizing state-of-the-art graphics cards as parallel processors for solving the nonlinear Schrödinger equation with the split-step Fourier method. Our parallel implementation of the fiber allows a speedup of up to a factor of more than 200 with comparable accuracy compared to current CPUs (shown in more detail in the next chapter).

References

1. Agrawal, G.P.: Nonlinear Fiber Optics, 3rd edn. Academic, San Diego (2001)
2. Agrawal, G.P.: Fiber Optic Communication Systems. Wiley, New York (1992)
3. Sudo, S., Kawachi, M., Edahiro, T., Izawa, T., Shioda, T., Gotoh, H.: Low-OH-content optical fiber fabricated by vapour-phase axial-deposition method. IEE Electron. Lett. **14**(17), 534–535 (1978).
4. Voges, E., Petermann, K.: Optische Kommunikationstechnik. Springer, Berlin (2002)
5. Capmany, J., Pastor, D., Sales, S., Ortega, B.: Effects of fourth-order dispersion in very high-speed optical time-division multiplexed transmission. OSA Opt. Lett. **27**(11), 960–962 (2002).
6. Dochhan, A., Smolorz, S., Rohde, H., Rosenkranz, W.: Electronic equalization of FBG phase ripple distortions in 43 Gb/s WDM systems. In: 10th ITG Conference on Photonic Networks, Leipzig, Germany, pp. 175–181 May 2009
7. Alfiad, M.S., van den Borne, D.: 11 Gb/s transmission with compensation of FBG-induced phase ripple enabled by coherent detection and digital processing. In: European Conference on Optical Communication (ECOC), P4.10, Vienna, Sept 2009
8. Westhäuser, M., Remmersmann, C., Pachnicke, S., Johansson, B., Krummrich, P.: Optimization of Optical Equalization of Group Delay Ripple-Induced Penalties from Fiber Bragg Gratings in 112 Gbit/s Metro Networks, accepted for OSA Optics and Photonics Congress "Advanced Photonics", Karlsruhe, Germany, June 2010
9. Desurvire, E.: Erbium Doped Fiber Amplifiers: Principles and Applications. Wiley, New York (1994)
10. van den Borne, D., Hecker-Denschlag, N.E., Khoe, G.-D., de Waardt, H.: Cross phase modulation induced depolarization penalties in 2x10 Gbit/s polarization-multiplexed transmission. In: European Conference on Optical Communication (ECOC), Mo4.5.5, Stockholm, September 2004
11. Winter, M., Kroushkov, D., Petermann, K.: Polarization-multiplexed transmission system outage due to nonlinearity-induced depolarization. In: European Conference on Optical Communication (ECOC), Th.10.E.3, Torino, 2010
12. Pachnicke, S., Voges, E., De Man, E., Gottwald, E., Reichert, J.: Experimental investigation of XPM-induced birefringence in mixed-fiber transparent optical networks. In: Optical Fiber Communications Conference (OFC 2006), JThB9, Anaheim, USA, Mar 2006

13. Mecozzi, A., Clausen, C.B., Shtaif, M.: System impact of intra-channel nonlinear effects in highly dispersed optical pulse transmission. IEEE Photon. Techn. Lett. **12**(12), 1633–1635 (2000).
14. Essiambre, R.-J., Mikkelsen, B., Raybon, G.: Intra-channel cross-phase modulation and four-wave mixing in high-speed TDM systems. IEE El. Lett. **35**(18), 1576–1578 (1999).
15. Killey, R.I., Thiele, H.J., Mikhailov, V., Bayvel, P.: Reduction of intrachannel nonlinear distortion in 40-Gb/s-based WDM transmission over standard fiber. IEEE Photon. Techn. Lett. **12**, 1624–1626 (2000). December
16. Plura, M., Kissing, J., Gunkel, M., Lenge, J., Elbers, J.-P., Glingener, C., Schulz, D., Voges, E.: Improved split-step method for efficient fibre simulations. IEE Electron. Lett. **37**(5), 286–287 (2001).
17. Kremp, T., Freude, W.: Fast split-step wavelet collocation method for WDM system parameter optimization. IEEE/OSA J. Lightw. Technol. **23**(3), 1491–1502 (2005)
18. Hellerbrand, S., Hanik, N.: Fast implementation of the split-step fourier method using a graphics processing unit. In: Optical Fiber Communication Conference (OFC), OTuD7, Mar 2010
19. Plura, M., Kissing, J., Lenge, J., Schulz, D., Voges, E.: Analysis of an improved split-step algorithm for simulating optical transmission systems. AEÜ Int. J. Electron. Commun. **56**(6), 361–366 (2002)
20. Pachnicke, S., Chachaj, A., Remmersmann, C., Krummrich, P.: Fast parallelized simulation of 112 Gb/s CP-QPSK transmission systems using Stratified Monte-Carlo sampling. In: Optical Fiber Communications Conference (OFC 2011), Los Angeles, USA, Mar 2011
21. Pachnicke, S., Chachaj, A., Helf, M., Krummrich, P.: Fast parallel simulation of fiber optical communication systems accelerated by a graphics processing unit. In: IEEE International Conference on Transparent Optical Networks (ICTON 2010), Munich, Germany, July 2010
22. Bosco, G., Carena, A., Curri, V., Gaudino, R., Poggiolini, P.: Suppression of spurious tones in fiber systems simulations based on the split-step method. In: IEEE LEOS Annual Meeting, WH 4, San Francisco, Nov 1999
23. Francia, C.: Constant step-size analysis in numerical simulation for correct four-wave-mixing power evaluation in optical fiber transmission systems. IEEE Photon. Technol. Lett. **11**(1), 69–71 (1999)
24. Jones, R.C.: A new calculus for the treatment of optical systems. J. Opt. Soc. Amer. **31**, 488–493 (1941)
25. Stokes, G.G.: On the composition and resolution of streams of polarized light from different sources. Trans. Cambridge Phil. Soc. **9**, 399–416 (1852)
26. Gordon, J.P., Kogelnik, H.: PMD fundamentals: polarization mode dispersion in optical fiber. Proc. Nat. Academy Sci. **97**(9), 4541–4550 (2000)
27. Frigo, N.J.: A generalized geometrical representation of coupled mode theory. IEEE J. Quant. Electron. **22**(11), 2131–2140 (1986)
28. Windmann, M., Pachnicke, S., Voges, E.: PHOTOSS: the simulation tool for optical transmission systems. In: SPIE ITCOM, Orlando, USA, 51-60 Sept 2003
29. Bergano, N.S., Kerfoot, F.W., Davidson, C.R.: Margin Measurement in Optical Amplifier Systems. IEEE Photon. Technol. Lett. **5**(3), 304–306 (1993)
30. Essiambre, R.-J., Kramer, G., Winzer, P.J., Foschini, G.J., Goebel, B.: Capacity Limits of Optical Fiber Networks. IEEE/OSA J. Lightw. Technol. **28**(4), 662–701 (2010)
31. Steele, R.C., Walker, G.R., Walker, N.G.: Sensitivity of optically preamplified receivers with optical filtering. IEEE Photon. Technol. Lett. **3**(6), 545–547 (1991)
32. Gnauck, A.H., Winzer, P.J.: Optical phase-shift-keyed transmission. IEEE J. Lightw. Technol. **23**(1), 115–130 (2005)
33. Forestieri, E.: Evaluating the error probability in lightwave systems with chromatic dispersion, arbitrary pulse shape and pre- and postdetection filtering. IEEE J. Lightw. Technol. **18**(11), 1493–1503 (2000)

34. Lee, J.S., Shim, C.S.: Bit-error-rate analysis of optically preamplified receivers using an eigenfunction expansion method in optical frequency domain. IEEE J. Lightw. Technol. **12**(7), 1224–1225 (1994).
35. PHOTOSS—The Photonic System Simulator. http://www.photoss.de
36. Virtual Photonics, VPI transmission maker. http://www.vpiphotonics.com
37. RSoft Design Group, OptSim^{TM} http://www.rsoftdesign.com
38. Optiwave, OptiSystem—Optical Communication System Design Software. www.optiwave.com

Chapter 4
Efficient Design of Fiber Optical Transmission Systems

Abstract Different aspects of the design of fiber optical transmission systems are described in this chapter. First a multidimensional meta-model based optimization algorithm is presented. This algorithm allows finding the optimum system parameters in a large parameter space with a highly reduced number of numerical simulations. Afterwards the parallelization of numerical simulations based on the split-step Fourier method is discussed. It is shown that by the use of standard desktop PC graphics cards featuring a high number of processing cores, the computational time for a fiber optical transmission link can be reduced significantly compared to a simulation executed on a CPU. Apart from the analysis of the speedup, the accuracy of the results is investigated, and a method is presented assuring simulation results with a certain predefined maximum relative error. Finally, analytical and semi-analytical modeling of the signal quality is explained. It is shown that by analyzing the different degradation effects occurring along a transmission line separately in many cases an analytical assessment is possible. This facilitates an estimation of the signal quality in a very short time of only a few seconds.

Generally, the task of a system designer is to find a setup of a transmission system with the minimum amount of total ownership costs, however, meeting the requirements for network reliability during the entire lifetime of the system. This demands being robust against variations of certain system parameters. These variations may occur due to changes in the environment (e.g. differences in the ambient temperature in the course of the year or externally induced vibrations), due to aging of components or dynamic changes in the network traffic and unexpected failures. For finding the optimum setup of a transmission system simulations are a key element. These may either be based on numerical solving of the nonlinear Schrödinger equation by the split-step Fourier method (compare Chap. 3) or on analytical models. In the following subsections different approaches for reducing the computational time needed for the simulation of a fiber optical transmission system are presented.

All different approaches presented in this chapter have been compared to results from system simulations with realistic parameters and also partly to laboratory experiments. At the end of the chapter a summary and a discussion are included.

4.1 Meta-Heuristic Based Optimization

Finding the set of parameters that lead to the best performance of an optical transmission system is an important step in the design process. Modeling and simulation play a major role in today's research and development of optical networks. Numerical simulations for optical fiber links are typically based on the split-step Fourier method (SSFM), which unfortunately requires a high computational effort. Especially for transmission systems with a high number of wavelength division multiplex (WDM) channels and a large transmission distance, where fiber nonlinearity cannot be neglected, the computational time for a single parameter combination lies in the range of several hours or days on a state-of-the-art desktop computer, which is the standard equipment for a system designer.

To optimize the performance of a fiber optical transmission system in many cases a search in a large parameter space is needed. Often there exist several interdependencies between different system parameters so it is not possible to reduce the search space to a small region in advance. A multidimensional meta-model based optimization algorithm is a good candidate for solving such a problem. The meta-heuristic (based on our publication [1]) presented below comprises several individual steps. First, a sparse number of parameters is selected from the entire parameter space by Latin hypercube design. For these parameters numerical simulations of the simulation system are performed consecutively. Based on these results an interpolation function is computed and an optimization algorithm is executed on this function to locate optima. Iteratively the interpolation function is improved. Apart from localizing an optimum parameter combination also regions of interest can be identified. The reason is that if the optimum parameter combination lies very close to a poor system performance, which may be adopted easily by the expected fluctuations of parameters in a real system or imprecision in the production process, the optimum is of little value. For additional sampling points further numerical simulations are executed in the identified regions of interest, and the accuracy of the interpolation function is enhanced continuously. In this way an iterative meta-model based optimization method has been created, which selects further sampling points based on so far evaluated data and carries out only essential calculations.

In the past a number of papers have been published on the topic of optimized design of optical transmission systems (e.g. [2–5]). However, most papers have restrictions in the maximum number of parameters, which can be varied or derive engineering rules for specific system configurations based on extensive numerical simulations (to find e.g. the optimum pre-compensation value). Furthermore, for certain system configurations fast analytical models for the assessment of the signal quality have been found (e.g. [6, 7]), especially for the 10 Gb/s non-return to zero (NRZ) on–off keying (OOK) modulation format (compare also Sect. 4.3 of this book). Analytical models or engineering rules have also been published for higher channel bit rates of 40 Gb/s or 160 Gb/s (e.g. [8, 9]). For other bitrates and more sophisticated modulation formats (e.g. DQPSK, QAM, etc.) the models have to be adapted and so far SSFM-based numerical simulations are the best way of analyzing the signal quality accurately. A single simulation of a long haul transmission system can take as long as several hours or even days on a current desktop computer (e.g. 30 min for the simulation software *PHOTOSS* presented in [10] and a 10×10 Gb/s NRZ-OOK transmission system with 10×100 km spans and a launch power of 0 dBm into the transmission fiber on a Intel® Core™ 2 Quad Q6600 with 2.4 GHz clock rate and 2 GB RAM. The simulation bandwidth has been 5.12 THz, and the duration of the simulated bit sequence totaled 6400 ps). Because simulations for different parameter sets are independent from each other it is possible to reduce the total amount of time needed for scanning a large parameter set by parallelization in a computing grid. However, the total computational effort is limited and compromises have to be made. For some investigations, such as optimization of the dispersion map, it is not unusual to utilize a computer pool of a double digit number of machines for several weeks or months (compare e.g. [11]).

4.1.1 Overview of Employed Algorithms

A. Optimization Methods

Mathematically optimization is characterized as the process of finding an optimal set of parameters for a given function. Optimal means that a figure of merit (or function) is maximized or minimized. As an optimization algorithm local search [12] or simulated annealing [13] may be used. Simulated annealing is usually employed when the complexity of the problem does not allow evaluating all possible combinations of parameters. The idea is based on the simulation of a cooling process of a certain material. At high temperatures the molecules are barely bound to their initial position. A given arrangement of the molecules can be regarded as an energy state. States of lower energy are preferred, and a lower energy state is adopted, if possible. The temperature dependent movement of the molecules allows to leave local minima for a higher energy state and to reach better energy states from this point than in the previous local optimum. Simulated annealing can be regarded as a local search where a solution from the neighborhood may be selected, even if it is not the best of the regarded neighborhood or worse than the current solution. Because the chance of accepting a worse solution is decreasing with a lower temperature, simulated annealing converges to a local search and terminates in a local optimum. The probability that this local optimum is identical to the global optimum, however, is much higher than for ordinary local

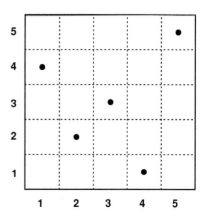

Fig. 4.1 Visualization of an LHD (5, 2) exemplary design

search. As an example in our model we used simulated annealing with the following parameters: maximum number of steps until the temperature is decreased = 50; initial temperature = 5; temperature factor = 0.7. The temperature factor is used to find the new temperature by multiplication with the current temperature. The probability that a new solution l_{try} is accepted is 1, if $f(l_{try}) < f(l)$, where $f(l)$ is the current solution and otherwise

$$P(f(l_{try}), f(l)) = \exp\left(-\frac{f(l_{try}) - f(l)}{t}\right) \quad (4.1\text{-}1)$$

where t is the current temperature and f the function to be optimized.

B. Design of Experiments

If a large region of a parameter space needs to be analyzed and obtaining each individual sampling point is costly (e.g. due to high computational or experimental effort), the question arises which sampling points should be evaluated. This problem is the same for simulations and experiments. Because each sampling point can be associated with a certain cost, a maximum number of sampling instances n can be defined as affordable. The problem of choosing the best sampling instances from the parameter space—if the maximum number of experiments (or simulations) is restricted—is referred to as design of experiments. The simplest way of distributing the sampling points is to use either a random distribution or an equidistant spacing. However, statistical methods, which make use of more intelligent approaches such as Latin hypercube design (LHD) have been proposed. LHD has been introduced by McKay, Conover and Beckman in 1979 [14].

A Latin hypercube is a generalization of a Latin square, which is a quadratic grid containing sampling points exactly once in each row and each column, for an arbitrary number of dimensions. An example of an LHD with the dimensions (5, 2) is given in Eq. 4.1-2 and depicted in Fig. 4.1.

4.1 Meta-Heuristic Based Optimization

$$LHD(5,2) = \begin{pmatrix} 1 & 4 \\ 5 & 5 \\ 3 & 3 \\ 4 & 1 \\ 2 & 2 \end{pmatrix} \qquad (4.1\text{-}2)$$

The advantage of distributing the samples with LHD compared to random distribution is that the position of the sampling points with regard to each other is considered by the algorithm. Thereby LHD guarantees that the sampling points are (more or less) equidistantly distributed with regard to any parameter, and no clustering occurs. For the two given values n and s there exist a multitude of possible LHD(n, s) designs. Typically one of these possible designs is selected randomly. Alternatively the so called maximin distance design can be used. A maximin design belongs to the class of distance related designs. It maximizes the minimal distance between every combination of two sampling points and guarantees that two points are not located too close together. In [15] a combination of LHD and maximin design is proposed yielding a distribution of the sampling points, which is equidistant in the given range of each parameter and also well distributed in all dimensions. A maximin LHD is obtained by successive modification of an LHD until the maximin property is achieved. Algorithmically this is done by using an optimization algorithm such as simulated annealing and a fitness function, which evaluates the maximin properties of the LHD. For a given LHD $\mathfrak{D}$ two lists can be computed: A distance list $d = (d_1,\dots,d_m)$, which contains the distances between the different sampling points in ascending order and a second list $J = (J_1,\dots,J_m)$, which contains for each distance in the first list the number of pairs of sampling points from $\mathfrak{D}$ that are separated by the respective distance. For the given class of LHD(n, s) designs a maximin fitness function can be derived by [15].

$$\Phi_p(\mathfrak{D}) = \left(\sum_{j=1}^{m} J_j d_j^{-1} \right)^{\frac{1}{p}} \qquad (4.1\text{-}3)$$

In the following example we selected the parameter p of the fitness function as $p = 5$.

C. Approximation and Interpolation Functions

Approximation or interpolation functions can be used to estimate the transmission quality at sampling points for which no results have been obtained (due to the sparse sampling of the parameter space). In this book scattered data interpolation is used (e.g. [16, 17]), which is suited to our problem very well because only a fraction of sampling points needs to be evaluated. In the following we present an example for an approximation function and the Hardy multiquadric interpolation function. The latter will be used extensively in the remainder of this section. The reason for choosing the Hardy multiquadric interpolation function from the large number of available scattered data interpolation methods is its ability to fit

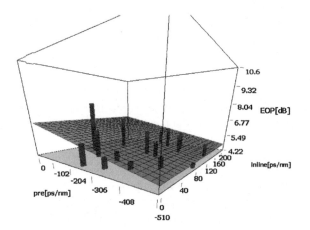

Fig. 4.2 Visualization of an equalization plane (*grid*) for the given sampling points (*bars*)

the underlying surface very accurately [18], which is especially important when an optimization is carried out on the interpolation function.

For a rough estimation of the unknown function values, a hyperplane can be computed, which lies as close as possible to the known sampling points (Fig. 4.2).

In general, interpolation through data points is desirable, if the values of the data points are highly accurate and reliable and if their number and density are not too high [18]. Interesting classes of interpolation functions are the radial basis functions [17]. The name already indicates that the interpolation function is generated in such a way that only the distance of the sampling point to an interpolation point is regarded. A scattered data interpolation function for a radial basis function can be defined as

$$f(x) = \sum_{i=1}^{N} \alpha_i R(d_i(x)) + p_m(x) \quad (4.1\text{-}4)$$

with

$$p_m(x) = \sum_{j=1}^{m} \beta_j \rho_j(x) \quad (4.1\text{-}5)$$

where $d_i(x)$ is the distance of point x to the i-th interpolation point, and $R(d_i(x))$ is a radial positive function. ρ_j are polynomials with a degree smaller or equal to m. The interpolation function depends on the coefficients α_i with $i = 1, .., N$ and β_j with $j = 1, .., m$. A system of equations is given by

$$\sum_{i=1}^{N} \alpha_i \rho_j(x_i) = 0, \quad j = 1 \ldots m, \quad (4.1\text{-}6)$$

which needs to be solved to find the coefficients. For multiquadric equations the parameter m is set to 0 yielding

4.1 Meta-Heuristic Based Optimization

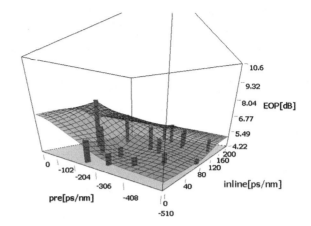

Fig. 4.3 Visualization of the Hardy multiquadric interpolation function with $\mu = 1$ (*grid*) for the given sampling points (*bars*)

$$f(x) = \sum_{i=1}^{N} \alpha_i R(d_i(x)). \qquad (4.1\text{-}7)$$

The basis function R for the Hardy multiquadric [17] is given by

$$R(r_i) = \left(r_i^2 + R_i^2\right)^{\frac{\mu_i}{2}}, \quad \mu \neq 0 \qquad (4.1\text{-}8)$$

where r_i, R_i and μ have to be chosen by the user. Usually r_i is set to $d_i(x)$. Furthermore, we selected $\mu = 1$ and $R_i = 0$. The extrema of the Hardy multiquadric interpolation function may not necessarily be equal to the extrema of the sampling points. The Hardy interpolation function thus allows continuing a trend. In the following examples the coefficients are calculated by a minimization problem to avoid numerical stability problems

$$\min_{\vec{a}} \left\| \begin{pmatrix} R(d_1(x_1)) & R(d_2(x_1)) & \cdots & R(d_n(x_1)) \\ R(d_1(x_2)) & R(d_2(x_2)) & \cdots & R(d_n(x_2)) \\ \vdots & \vdots & \vdots & \vdots \\ R(d_1(x_n)) & R(d_2(x_n)) & \cdots & R(d_n(x_n)) \end{pmatrix} \begin{pmatrix} \alpha_1 \\ \alpha_2 \\ \vdots \\ \alpha_n \end{pmatrix} - \begin{pmatrix} f_1 \\ f_2 \\ \vdots \\ f_n \end{pmatrix} \right\| \qquad (4.1\text{-}9)$$

where (x_i, f_i) are the n known sampling points. The resulting interpolation function is depicted exemplarily in Fig. 4.3.

4.1.2 Meta-Model

The goal of the proposed meta-model is to optimize the parameters of an optical transmission system with carrying out as few numerical simulations of the transmission system as possible. However, it has to be kept in mind that only in simulations the parameters can be defined precisely. In a real system the parameters may fluctuate or differ from the simulated parameter set (due to inherent tolerances of the production process, other varying external conditions or finite granularities of certain component parameters). The optimum parameter combination is of little value, if the optimum lies very close to a poor system performance, which may be reached easily by the expected fluctuations of parameters in a real system. These are the reasons why not only a single optimum parameter combination should be searched for, but an optimal region with a defined width. The quality of such a region can be evaluated by the average of all values in this region or the worst-case value occurring in it. In the following the worst-case value (of the respective region) is taken as the criterion. Because the fluctuation of the individual parameters is assumed to be independent from each other, the width of the area can be defined for each parameter independently. In other words a region is searched for, which is limited in each dimension by an upper and a lower bound. This region is referred to as a hyper cuboidal subset $\mathfrak{L}$ of the d-dimensional parameter space. The set of all possible subsets of the parameter space is termed $\mathfrak{L}(\vec{s})$. From this set the subset $\mathfrak{L}$ is searched for, which is optimum (in our case regarding the worst-case performance).

To identify such a region, a meta-heuristic has been defined. For this heuristic the following parameters are needed: a function f to be optimized, the minimal width s_i of the interval for each parameter and upper ($\vec{o}$) and lower ($\vec{u}$) boundaries of the initial search region. Of course the following inequality has to be fulfilled:

$$o_i - u_i \geq s_i. \tag{4.1-10}$$

In the initial region the search is started, and the upper and lower boundaries are moved, and the distance is decreased until the exit condition is reached:

$$\forall i, 1 \leq i \leq d : (o_i - u_i = s_i). \tag{4.1-11}$$

The area, which is ultimately bounded by $\vec{o}$ and $\vec{u}$ is the result of the optimization meta-heuristic. A pseudo-code of the algorithm is shown in the listing below.

Algorithm 1: Meta-model based optimization algorithm

1. Set $\vec{o}, \vec{u}$ to the values of initial region
2. **repeat**
3. $\vec{d}$ = determine_sampling_points_in_region $(\vec{o}, \vec{u})$
4. $\vec{y} = f(\vec{d})$
5.   model = update_model_with_new_values (model, $\vec{d}, \vec{y}$)
6. $\vec{v}$ = optimum_region (model)
7. $\vec{o}, \vec{u}$ = update_search_area $(\vec{v})$
8. **until** $\forall i : o_i - u_i = s_i$

The function *determine_sampling_points_in_region* returns a given number of sampling points in the defined region $(\vec{o}, \vec{u})$. Either LHD or maximin LHD are employed. In line 4 the function is evaluated at the sampling points determined before. In practice this means that numerical simulations for the transmission

4.1 Meta-Heuristic Based Optimization

system with the given parameter combinations are executed. Afterwards the model is refined with the newly calculated results from the numerical simulation. Algorithmically this is done by Hardy-multiquadric interpolation outlined above. In line 6 the *optimum_region* is searched for. In our case a simulated annealing algorithm is executed. Finally, the search area is updated. The pseudo-code for this calculation is listed below.

Algorithm 2: update_search_area $(\vec{v})$

1. $t_i = o_i - u_i$ (for all i)
2. $t_i = t_i \cdot fac_i$ (for all i)
3. **if** $t_i < s_i$ **then**
4. $\quad\quad t_i = s_i$
5. **end if**
6. $\vec{o} = \vec{v} + \vec{t}/2$
7. $\vec{u} = \vec{v} - \vec{t}/2$

In practice algorithm 2 reduces the range of the search area in each step by a factor fac_i ($0 < fac_i < 1$). fac_i must be defined by the user. If the factor fac_i is reaching 1, the total number of steps approaches infinity. On the other hand, if $fac_i \rightarrow 0$, the total number of steps approaches 1. In line 3 the function *determine_sampling_points_in_region* is called. The higher the number of sampling points, which are returned by the function, the more accurate is the approximation of the function. Regarding the computational time, however, this parameter should be as small as possible. If the underlying (unknown) function is expected to change rapidly, the number of sampling points and the factor *fac* should be increased. Because our algorithm is not searching for a single optimum parameter combination, but an optimum region, however, the meta-model is relatively insensitive to more rapid changes of the graph. In the following section the optimum choice of parameters is discussed.

4.1.3 Analysis of Exemplary Transmission Systems

The meta-heuristic has been tested for two transmission system examples. The setup is shown in Fig. 4.4. The first setup is a 10 × 10 Gb/s NRZ-OOK WDM transmission system with 5 spans of 100 km each on a 50 GHz channel grid. The system setup and transmission distance have been kept simple to reduce the total computational time needed for the grid search, which has been used as a benchmark. This is also why in all simulations a pseudo-random bit sequence (PRBS) length of 2^6-1 and 512 samples per bit have been used. The results obtained for a total of 12,240 parameter combinations serve as a reference for the meta-heuristic.

In the simulations the dispersion map (pre-DCF, inline-DCF and post-DCF) as well as the launch power have been varied. The parameters of the optical and

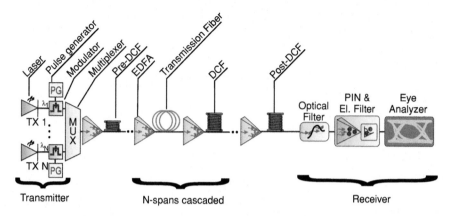

Fig. 4.4 Setup of the simulated transmission system

electrical filters have been assumed fixed ($\Delta f_{opt} = 40$ GHz and $\Delta f_{el} = 7$ GHz). Standard single mode fiber (SSMF) has been used as a transmission fiber. The pre-DCF has been varied between −350 ps/nm and 0 ps/nm (in 25 ps/nm steps), the inline-DCF between 0 and 165 ps/nm residual dispersion (per span, in 15 ps/nm steps), and the post-DCF allows to set the residual dispersion at the receiver between 0 and 960 ps/nm (in 60 ps/nm steps). For the launch power a value between -1 and 2 dBm per channel has been chosen.

As a second system a metro system with 10×107 Gb/s transmission system in a 100 GHz channel grid has been analyzed. Again 5 spans with 100 km each have been used. As a modulation format this time duobinary modulation has been chosen. The filter parameters are $\Delta f_{opt} = 100$ GHz and $\Delta f_{el} = 75$ GHz. The other parameters are as follows: pre-DCF between -510 and 0 ps/nm (in 10 ps/nm steps), inline DCF between 0 and 200 ps/nm (in 8 ps/nm steps) and launch power between -1 and 2 dBm per channel. The post-DCF has been used to perfectly compensate for the residual dispersion. A total of 5,408 parameter combinations have been simulated in this case. In all simulations the eye opening penalty (EOP) of the center channel has been used as a figure of merit. ASE noise has been simulated analytically [10] and is not included in the EOP.

First, the accuracy of the interpolation functions has been investigated. As a figure of merit the relative error of the functational fit with regard to the sampling points (evaluated by the SSFM) has been taken (normalized to the value of the respective sampling point). The relative error has been computed for each sampling point and subsequently averaged over all sampling points (12,240 for the 10×10 Gb/s and 5,408 for the 10×107 Gb/s transmission system). This investigation has been repeated for 100 different realizations of maximin LHD (and averaged over all realizations).

In Fig. 4.5 the number of sampling points has been varied between 5 and 195. In all figures also the 90 % confidence intervals are depicted. For each point

4.1 Meta-Heuristic Based Optimization

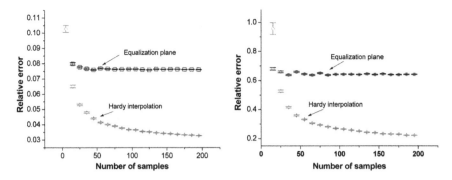

Fig. 4.5 Comparison of the accuracy of different functional fits for the 10 × 107 Gb/s transmission system (left) and the 10 × 10 Gb/s transmission system (*right*)

(i.e. number of sampling points) 100 simulations have been carried out with different positions of the sampling points determined by maximin LHD. When comparing the left and right graph in Fig. 4.5 it turns out that in both cases the Hardy interpolation method outperforms the equalization plane.

In further investigations it has been studied how the optimization algorithm performs in combination with the Hardy interpolation method. Again the number of sampling points has been varied. The simulated annealing algorithm has been used to find the optima on the interpolation functions. As a figure of merit this time a relative deviation from the real optimum has been used given by

$$Dev = \frac{f(o_{est}) - f(o_{real})}{f_{max} - f_{min}} \qquad (4.1\text{-}12)$$

where o_{est} and o_{real} are the optimal sampling points determined by the simulated annealing algorithm and the real optimum, respectively. f_{max} is the maximum function value and f_{min} the minimum value. Here the function values reflect the EOP (in decibels).

As a reference also the performance of the simulated annealing algorithm based on only maximin LHD sampling points is depicted in Fig. 4.6. Clearly the use of the Hardy interpolation method shows an outperformance. Furthermore, it can be seen that increasing the number of sampling points to more than 35 does not offer a significant advantage.

To evaluate the performance of the proposed meta-heuristic a search for the optimum region for the 10 × 107 Gb/s transmission system has been conducted. For the region of interest a width of 30 ps/nm for the pre-DCF and of 24 ps/nm for the inline DCF has been set. In each iteration of the meta-heuristic two sampling points have been evaluated.

The ability of the meta-heuristic to find the optimum region is investigated. As a figure of merit the ratio of the optimum identified by the meta-heuristic to the optimum determined by a grid search has been used. The factor *fac* determines with which factor the search area is multiplied for the next iteration (thus a larger

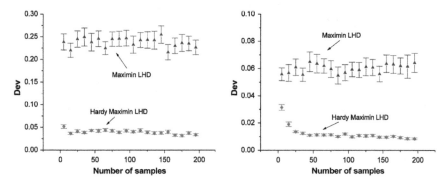

Fig. 4.6 Relative deviation from the optimum determined by simulated annealing to the real optimum for the 10×107 Gb/s transmission system (left) and the 10×10 Gb/s transmission system (right)

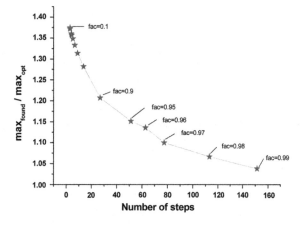

Fig. 4.7 Comparison of deviation from the optimum for the Hardy interpolation method and different factors fac

fac leads to a smaller decrease of the area in each iteration). The total number of steps needed for different fac and the resulting deviation from the optimum are depicted in Fig. 4.7. As an example for a factor $fac = 0.9$ and 2 sampling points per iteration it is only necessary to evaluate $2 \cdot 25$ sampling points out of the total of 5,408 evaluated in standard grid search with only a marginally less accurate solution than the optimum. If the number of sampling points per iteration is doubled (leading to a total of 100 evaluated sampling points), the identified optimum region is almost identical to the optimum determined by the grid search.

To show that our meta-model based heuristic can handle more parameters than other approaches we set up a simulation with a high number of variables. The setup is based on a 10.7 Gb/s NRZ-OOK WDM transmission system with 10 SSMF spans of 80 km each on a 50 GHz grid. All EDFAs had a noise figure of 5.5. dB. As a figure of merit the bit error ratio of the center channel has been assessed. The varied parameters are the DCF lengths of all 10 DCFs, the pre-DCF, the post-DCF, the launch power into the transmission fibers and the launch power

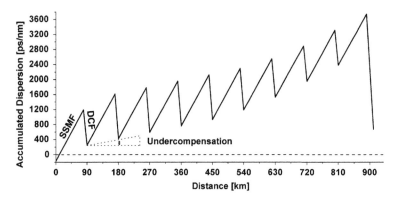

Fig. 4.8 Dispersion map determined by the meta-heuristic

into the DCF modules making up a total of 14 independent variables. In the simulations a PRBS length of 2^6-1 and 128 samples per bit have been used.

The dispersion map, which has been determined by the meta-heuristic, is depicted in Fig. 4.8. For the DCF modules a granularity of 85 ps/nm has been set. The computed optimum dispersion map begins with a pre-compensation of 170 ps/nm and ends with a residual dispersion at the receiver of 680 ps/nm. After the first span the algorithm suggests to use a shorter DCF leaving a residual dispersion of 255 ps/nm. Also for spans 8 and 9 shorter DCFs have been suggested. The meta-heuristic initially searched in the range of −340 ps/nm (overcompensation) to 510 ps/nm (undercompensation) in each span. For the pre-DCF a range of −850 to 0 ps/nm has been set initially. The launch powers could vary between −1 and 3 dBm/ch for the SSMF and between −2 and 2 dBm/ch for the DCF. The meta-heuristic determined 0 dBm/ch into the transmission fiber and −2 dBm/ch into the DCF as optimum. The results are similar to other results from the literature (e.g. [19]), however, our meta-heuristic enables to find these values in the large 14-dimensional parameter space with computing only a very low number of 82 numerical (SSFM-based) simulations.

4.2 Parallelization of a Simulation on a Graphics Processing Unit

The meta-heuristic based optimization algorithm shown in the last section makes use of numerical simulations of the fiber optical transmission system, which are based on the split-step Fourier method (SSFM). These simulations unfortunately exhibit a relatively high computational complexity (and related computational time) for the desired accuracy of the results. Especially for transmission systems with significant nonlinear fiber degradation effects, a large number of WDM channels and a very long transmission distance the computational time can take several days or weeks on a state-of-the-art desktop PC (e.g. Intel® Core™ i7), which is typically used by a system designer today.

A possible way of speeding up these numerical simulations is to use parallelization techniques, which are addressed in this section. Especially the FFT and IFFT operations required in the SSFM can be parallelized efficiently promising a high speedup on a parallel computer. Today, graphics cards offer a high number of processing cores at a reasonable cost and are already available in many standard desktop PCs. This is why we have investigated graphics cards as a potential candidate for speeding up the numerical simulations. Current graphics cards are mainly designed for gaming purposes, which do not require high accuracy calculations. This is why they offer a much higher computational power in single precision (SP) accuracy than in double precision (DP) accuracy. For scientific computing NVIDIA® has recently announced a special series of graphics cards called Tesla™ Fermi™. These graphics cards have an improved DP performance, but still SP computing power is twice as high as in DP mode. For these reasons SP based simulations are desirable. Unfortunately, the intrinsic FFT implementation (CUFFT) provided with NVIDIA® graphics cards used in this book shows a relatively high inaccuracy in SP mode making it unsuitable for many simulations with a high number of split-steps. This is why we present a novel implementation of the FFT in SP mode leading to a much higher accuracy than CUFFT at a simulation speed, which is not significantly reduced. Based on this new implementation, numerical simulations of optical transmission systems have been conducted with the SSFM, showing that it yields low inaccuracies compared to CPU based simulations in DP up to more than 100,000 split-steps. Some simulation systems with 80 WDM channels and very long transmission distances, however, may require an even higher number of several 100,000 split-steps. To ensure accuracy comparable to DP simulations also for these extreme cases, a stratified MC sampling method is proposed. With this method we have conducted transmission system simulations with up to 880,000 split-steps still showing comparable accuracy to DP simulations at a high speedup.

4.2.1 Implementation of the FFT and Split-Step Fourier Method on a GPU

For the implementation of the SSFM on graphics cards the general purpose computation on graphics processing units (GPGPU) framework Compute Unified Device Architecture (CUDA™) distributed with NVIDIA® graphics cards has been used. A CUDA™ program consists of several parts, which are executed on either the CPU or the graphics processing unit (GPU). The NVIDIA® C Compiler (NVCC) passes so called host code to the ANSI C compiler for the CPU and so called device code to a special compiler for the GPU.

NVIDIA® GPUs are multi-core (vector) processors based on the SIMD (single instruction multiple data) principle. The architecture is specially designed for a high speed of parallel computations, whereas the speed of a single thread and the

4.2 Parallelization of a Simulation on a Graphics Processing Unit 69

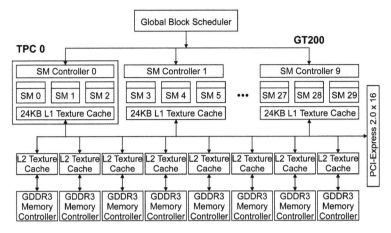

Fig. 4.9 NVIDIA® GT200 processor architecture (adapted from [20])

latency are of minor importance. This implies that there are always enough waiting tasks in the pipeline. Consequently no speculative execution is necessary in contrast to CPUs. In this book an NVIDIA® GTX 275 has been used, which has a total of 240 streaming processors. Its processor architecture (the GT200 architecture) is depicted in Fig. 4.9.

The GT200 consists of ten thread processing clusters (TPC), containing three streaming multiprocessors (SMs) and a texture pipeline, which is used as memory pipeline (shared between the three SMs). Each SM comprises eight streaming processors (used for 32 bit SP accuracy calculations) and one 64 bit DP unit. In SP the floating point accuracy is mostly IEEE-754 compliant (except for fused multiply and add operations), DP is entirely IEEE-754 compliant. However, the trigonometric functions, which are implemented in hardware in a so called special function unit, have a rather low accuracy in SP. Details of the NVIDIA® GTX 275 architecture are exposed in [21].

A. Implementation of the FFT on a GPU

In the following, the basics of the fast Fourier transform (FFT) and its parallelization shall be outlined. Most of the currently used FFT algorithms are based on the work by Cooley and Tukey [22]. To understand the theory behind the Cooley-Tukey algorithm, first the discrete Fourier transform (DFT) shall be defined:

$$V(\mu) = DFT\{v(k)\} = \sum_{k=0}^{M-1} v(k) \cdot w_M^{\mu k}, \quad \mu = 0, \ldots, M-1 \quad (4.2\text{-}1)$$

with

$$w_M^{\mu k} = \exp\left(\frac{-2\pi j \mu k}{M}\right) \quad (4.2\text{-}2)$$

being the so-called twiddle factor. $v(k)$ is the (complex) input sequence of length M. To solve Eq. 4.2-1 M^2 complex multiplications and the same number of additions are needed.

The idea of the FFT is to split up the input sequence into two sequences with even and odd arguments. For this purpose the input sequence must be a power of two ($M = 2^m$).

$$V(\mu) = \sum_{l=0}^{(M/2)-1} v(2l) \cdot w_M^{\mu \cdot 2l} + \sum_{l=0}^{(M/2)-1} v(2l+1) \cdot w_M^{\mu \cdot (2l+1)}, \quad \mu = 0, \ldots, M-1$$

(4.2-3)

With the help of the relation $w_M^{\mu \cdot 2l} = w_M^{\mu l}$ Eq. 4.2-3 can be rewritten yielding

$$V(\mu) = \sum_{l=0}^{(M/2)-1} v(2l) \cdot w_M^{\mu \cdot 2l} + w_M^{\mu} \cdot \sum_{l=0}^{(M/2)-1} v(2l+1) \cdot w_M^{\mu l}, \quad \mu = 0, \ldots, M-1$$

$$= V_1(\mu) + w_M^{\mu} \cdot V_2(\mu).$$

(4.2-4)

In this way the number of complex multiplications is reduced to $2(M/2)^2 + M$. Finally, in the $(m-1)$th step, $M/2$ DFTs of length 2 remain, which can be computed easily by

$$G(\mu) = \sum_{k=0}^{1} g(k) \cdot w_2^{\mu k}, \mu = 0, 1 \rightarrow \begin{array}{l} G(0) = g(0) + g(1) \\ G(1) = g(0) - g(1) \end{array}.$$

(4.2-5)

The Cooley-Tukey algorithm has a complexity of approximately $5n \cdot \log_2 n$ operations for an FFT length of 2^n.

To parallelize the above algorithm the input vector $v(k)$ is reorganized as a two-dimensional array with length $M = M_1 \cdot M_2$. If the definition of the DFT in Eq. 4.2-1 is kept, the indices have to be rewritten:

$$\begin{array}{ll} k = k_1 \cdot M_2 + k_2, & k_1 = 0, \ldots, M_1 - 1, \quad k_2 = 0, \ldots, M_2 - 1 \\ \mu = \mu_1 + \mu_2 \cdot M_1, & \mu_1 = 0, \ldots, M_1 - 1, \quad \mu_2 = 0, \ldots, M_2 - 1 \end{array}$$

(4.2-6)

yielding

$$V(\mu_1 + \mu_2 \cdot M_1) = \sum_{k_1=0}^{M_1-1} \sum_{k_2=0}^{M_2-1} v(k_1 \cdot M_2 + k_2) \cdot w_{M_1 \cdot M_2}^{(k_1 \cdot M_2 + k_2) \cdot (\mu_1 + \mu_2 \cdot M_1)}$$

$$= \sum_{k_1=0}^{M_1-1} \sum_{k_2=0}^{M_2-1} v(k_1 \cdot M_2 + k_2) \cdot w_{M_1}^{k_1 \cdot \mu_1} \cdot \underbrace{w_1^{k_1 \cdot \mu_2}}_{=1} \cdot w_M^{k_2 \cdot \mu_1} \cdot w_{M_2}^{k_2 \cdot \mu_2}$$

$$= \sum_{k_2=0}^{M_2-1} \left[\sum_{k_1=0}^{M_1-1} \left(v(k_1 \cdot M_2 + k_2) \cdot w_{M_1}^{k_1 \cdot \mu_1} \right) \cdot w_M^{k_2 \cdot \mu_1} \right] \cdot w_{M_2}^{k_2 \cdot \mu_2}.$$

(4.2-7)

Fig. 4.10 Parallelization of an FFT of length 16 into 8 FFTs of length 2 along the rows and 2 FFTs of length 8 along the columns

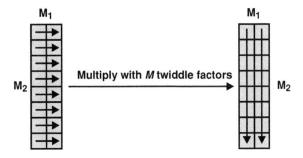

Fig. 4.11 Parallelization of an FFT of length 16 into 8 FFTs of length 2 along the rows and 2 FFTs of length 8 along the columns

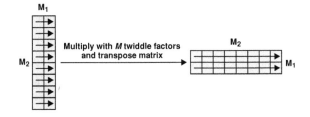

The algorithm calculates M_2 DFTs of length M_1 in the inner sum, multiplies the result with the twiddle factors $w_M^{k_2 \cdot \mu_1}$ and finally computes M_1 DFTs of length M_2 (Fig. 4.10). Similar to the 1D DFT shown above, the DFT in Eq. 4.2-7 can be split-up recursively until DFTs of length two (also called radix-2) are reached.

Instead of breaking down the DFT into smaller pieces to a radix-2 butterfly, it is also possible to implement higher radix sizes R directly. A specific radix size R generally requires the input sequence to be a power of R. Higher radix sizes have the advantage of requiring less complex multiplications and less stages. For an FFT length of e.g. 32,768 and a radix of two, 15 stages have to be passed (2^{15}), whereas for a radix of 32 only three stages would be required (32^3). Usually a maximum length of either M_1 or M_2 is given corresponding to the maximum available radix size, and in the other dimension the DFT is split up (if necessary). FFT algorithms, which use different sizes of the radices, are denoted as mixed radix algorithms.

It is in many cases advantageous that the input data is available in a continuous fashion. This is why a matrix transposition step may be introduced (Fig. 4.11). A further advantage of this transposition is that the results are available in their original sequence, and no additional digit reversal phase is needed. For the matrix transposition step an out-of-place algorithm can be used. This requires twice the memory for the input vector, however, facilitating a higher computational speed.

The above algorithm is called four-step-FFT-algorithm. It is obvious that all four steps can be parallelized.

It is difficult to predict the optimum choice of the radix on a GPU. The reason is that the performance strongly depends on the hardware architecture and the clock rate of the different memories and caches [23]. This is why we used an auto-tuning algorithm to determine the optimum configuration. We implemented two different kernels for each radix size. The first one computes the FFT, multiplies with the

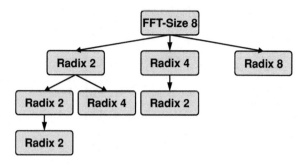

Fig. 4.12 Visualization of the different radix choices for an FFT length of 8

twiddle factors and does a matrix transposition. The second kernel realization computes an FFT only (compare Fig. 4.11, right). To optimize the memory access also kernels are available, which reorganize the data in the shared memory before writing into the global memory. This is needed, if a memory stride occurs which is not a multiple of 16, to avoid a (time-consuming) serialization of the memory accesses. The auto tuning algorithms determines the computational time for each kernel once for a given GPU hardware architecture. The optimum factorization of an FFT of a certain length can afterwards be calculated from a tree-like structure (Fig. 4.12). In such a tree all possible factorizations of an FFT with a certain length are depicted. Associated to each radix size is its computational time, which has been predetermined before. To find the optimum factorization, the path from a leaf to the root has to be traced in the tree and the associated computational times have to be added. The path with the lowest aggregate computational time is used.

The results obtained for an NVIDIA® GeForce® GTX 275 are depicted in Table 4.1.

Apart from the optimum speed of an FFT, also the accuracy of the results is crucial. GPU based simulations in DP accuracy do not show a largely increased error compared to CPU based simulations (more details can be found in the following section B). Because of the much higher computational power in SP accuracy, however, SP-based simulations are desired. As mentioned before, though, the CUDA^{TM} SP FFT realization (CUFFT) shows a rather high inaccuracy. The reasons for this behavior are twofold. First, the fused multiply and add (FMAD) units due not fulfill the IEEE-754 specifications. Second, the intrinsic (hardware implemented) trigonometric functions are rather inaccurate. This is especially detrimental for the implementation of the SSFM with a high number of split-steps as for each step one FFT and one IFFT is needed.

To increase the accuracy of the results thus the FMAD operations have been replaced by a more accurate version being compliant with the IEEE-754 specifications. This is easily possible because both multiplications and additions are implemented in an accurate way on the GPU, and only the FMAD operations show a higher inaccuracy. Furthermore, the trigonometric functions have to be improved regarding the accuracy. For this purpose it has been proposed to compute the twiddle factors in DP [24]. This requires calculating a Taylor series expansion in software as no fast hardware based solution is available on the employed GPUs. To avoid a high

4.2 Parallelization of a Simulation on a Graphics Processing Unit

Table 4.1 Factorization of our FFT-Implementation for different FFT-Lengths on NVIDIA® GeForce® GTX 275

FFT-Length	2^{12}	2^{13}	2^{14}	2^{15}	2^{16}	2^{17}	2^{18}	2^{19}	2^{20}
Our FFT	64×64	$16 \times 16 \times 32$	$16 \times 16 \times 64$	$16 \times 32 \times 64$	$32 \times 32 \times 64$	$64 \times 32 \times 64$	$64 \times 64 \times 64$	$64 \times 64 \times 128$	$64 \times 128 \times 128$

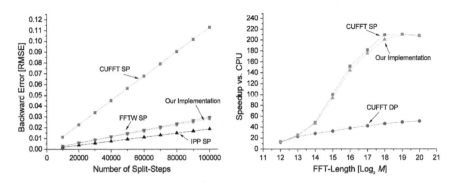

Fig. 4.13 Accumulation of the backward error vs. the number of split-steps for an FFT-length of 2^{20} (*left*) and speedup compared to a single CPU core (*right*)

reduction of the computational speed of the FFT an alternative is to pre-compute the twiddle factors with DP and to store them as a lookup table. This is especially advantageous as the lookup table can be reused during the course of the SSFM where thousands of FFTs are typically needed. In our implementation we chose to pre-compute the twiddle factors in DP on the CPU and to store them in the global graphics memory. Two different kinds of twiddle factors can be distinguished: local and global ones. The former are needed during the computation of the radices, and the latter ones in the multiplication step of the four-step FFT. Local twiddle factors are only dependent on the size of the radix and do not consume much memory space. They can be stored in the constant memory of the GPU (which is limited to 64 kB on the GT200). The advantage of this memory is its very high speed. Because for the global twiddle factors much more memory is needed (and also utilization of symmetry properties does not reduce the size below the 64 kB limit), only the global memory can be used for this purpose. We employed a separate lookup table for each factorization of the FFT because in this way a consecutive memory access onto the table can be guaranteed (consuming some more memory, however).

We used the backward error shown in Fig. 4.13 (left) to compare the accuracy of four different FFT implementations. The backward error is calculated by performing an FFT and IFFT operation for an arbitrary input sequence and comparing the results to the initial sequence. For the simulations we used CUDA runtime version 3.0. It can be observed that the backward error for the NVIDIA® CUFFT library in single precision is increasing steeply with the number of split-steps, whereas our implementation shows comparable accuracy to the Intel® IPP library (version 6.1) executed on a CPU and even exhibits a smaller error than the widely used FFTW library (version 3.2.2 in single precision). In double precision mode GPU-based results are almost identical to CPU-based results (compare [25–27]). In Fig. 4.13 (right) the results for the speedup compared to the same setup simulated on a CPU in double precision are depicted. It can be observed that in double precision mode GPU-based simulations on the NVIDIA® GTX 275 are up to 50 times faster than on our reference CPU Intel® Quad Core Q6600 (@2.4 GHz with 2 GiB RAM). The increased performance for a higher FFT length can be attributed to a better utilization of the GPU hardware. Furthermore the transfer between main memory and graphics memory is dominating the calculation time for smaller FFT sizes. In single precision arithmetic the speedup is rising to more than a factor of 210 as the number of single precision units on the GPU is significantly higher. Our own implementation is almost as fast as the NVIDIA® CUFFT library, however, at a much higher accuracy. Furthermore, it is interesting to note that for simulations with two polarization axes (as used for CP-QPSK) a batch FFT can be used concatenating two FFTs of the same length and leading to a performance of a FFT with double length, which is beneficial on a GPU in most cases.

B. Implementation of the Split-Step Fourier Method on a GPU

The signal propagation in optical fibers can be described by the nonlinear Schrödinger equation (NLSE), which has been presented in Chap. 3 in detail. A common way to solve the NLSE is to use the split-step Fourier Method (SSFM). In Sect. 3.1.6 the SSFM has already been described in detail. At this point it shall only be recalled that the idea of the SSFM is to divide the NLSE into a linear and a nonlinear part. The former one is solved in the frequency domain and the latter one in the time domain. The transformation between time and frequency domains and vice versa is obtained by FFTs and IFFTs, respectively. For these FFT/IFFT operations a parallelized implementation on a GPU can be used as explained above. At this point it shall also be mentioned that apart from the calculation of the FFTs (or IFFTs) also the arithmetic operations of the linear and nonlinear operators of the SSFM can be efficiently parallelized as they are executed on the entire length of the sampled signal vector, which is a SIMD operation.

To speedup the entire SSFM it is important to reduce the number of data transfers between host (CPU) and GPU, which are rather time consuming. In our implementation we only need two memory transfers: one at the beginning of the calculation of a fiber and another one at the end.

For the efficiency of the SSFM, the right choice of the step-size is crucial. If the step-size is chosen too large, the results will be inaccurate due to insufficient consideration of the interaction of the linear and nonlinear operators. On the other hand, if the step-size is chosen too small, the computational efficiency is decreased. Details have been presented in Sect. 3.1.6. In the following we used a conservative value of −50 dB of maximum allowed artificial FWM.

For the assessment of the accuracy of the simulations we selected the following reference scenario (Fig. 4.14).

The transmission system operated at a line rate of 10.7 Gb/s (NRZ-OOK) on a 50 GHz grid as an example for long-haul WDM transmission with considerable impact of nonlinear interchannel effects. We deployed a pre-DCF with −650 ps/nm. A distributed undercompensation scheme has been used along the transmission line with 50 ps/nm residual dispersion per span. At the receiver the accumulated dispersion has been driven to zero. SSMF has been assumed as transmission fiber. The launch power has been varied between −6 and 4 dBm/channel. The EDFAs in our simulations had a noise figure of 5.5 dB. We used a 50 GHz 2nd order Gaussian optical bandpass filter to select the desired channel and a 7 GHz 10th order

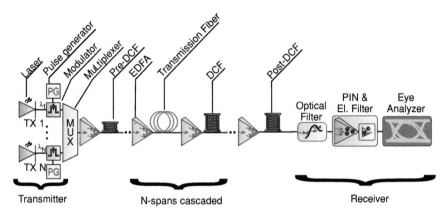

Fig. 4.14 Setup of the simulated transmission system

electrical Bessel filter. 10 WDM channels and 25 fiber spans of 80 km length each have been simulated. In all simulations 64 samples/bit have been used.

As a figure of merit we calculated the Q-factor of the center channel for the different setups and the two different simulation methods (GPU- and CPU-based). An analytical noise representation [10] has been used separating numerical simulation of the signal and analytical representation of the noise variance (compare also Sect. 3.1.8 of this book). At the receiver a Q-factor is derived assuming Gaussian distribution of the noise. Other noise sources—apart from ASE—such as lasers or photodiodes have been ignored.

In our studies we investigated setups with different numbers of spans and different FFT sizes (respectively PRBS lengths) starting with a double precision (DP) implementation based on the CUDA^{TM} intrinsic CUFFT library. We defined the Q-factor deviation ΔQ as the absolute value of the difference between CPU-based simulations and GPU-based simulations using the same random generator seed thus theoretically leading to identical results on the same hardware architecture.

$$\Delta Q \text{ [dB]} = \left| 20 \cdot \log_{10}\left(Q_{CPU, Double\,Prec}\right) - 20 \cdot \log_{10}(Q_{GPU}) \right| \quad (4.2\text{-}8)$$

The Q-factor deviation has been investigated at different span counts for a fixed launch power of 0 dBm/ch (Fig. 4.15, left) and for different launch powers at a given transmission distance of 20 spans (Fig. 4.15, right). As a reference the simulated Q-factor has been approximately 11.3 dB for a launch power of 1 dBm/ch and a transmission distance of 20 spans. As can be seen from Fig. 4.15 (left and right) the Q-factor difference between the CPU- and GPU-based simulations is negligible for all investigated FFT-lengths ($\Delta Q < 10^{-9}$ dB).

The depicted Q-factor deviations in Fig. 4.15 (left) start at a span count of 4 because for a lower number of transmission spans the signal quality is almost ideal, and no Q-factor could be determined (error free transmission). Fig. 4.15 (left) shows that higher FFT lengths do not necessarily lead to a (significantly) higher Q-factor deviation. However, it can be observed that the Q-factor deviation

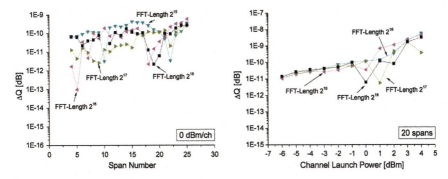

Fig. 4.15 Q-factor deviation between CPU and GPU-based simulations for different FFT lengths vs. the transmission distance in spans (*left*) and vs. the channel launch power at a given transmission distance (*right*). Both figures show results obtained in double precision mode

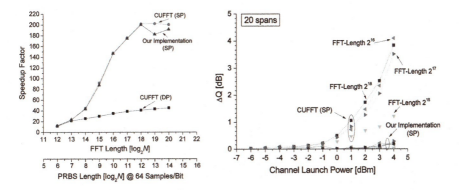

Fig. 4.16 Speedup factor for GPU-based compared to CPU-based simulations in double precision on a single core (*left*) and Q-factor deviation between CPU and GPU-based simulations in single precision for different FFT lengths vs. channel launch power (*right*)

increases with the span count, which is due to the increasing number of split-steps (or FFT/IFFT operations). To prove that the observed Q-factor deviation stems from the FFT/IFFT operation, we used a mixed GPU/CPU implementation computing the FFT/IFFTs on the CPU and the other operations on the GPU showing that the error vanishes for that configuration.

Another analysis is shown in Fig. 4.15 (right). Here the transmission distance has been fixed, and the channel launch power in each transmission fiber has been varied. It can be observed that the average Q-factor deviation increases from approximately 10^{-11} dB for a launch power of –6 dBm/ch to approximately 10^{-9} dB for a launch power of +4 dBm/ch. This is in line with the increase in the number of split-steps (504 per span for –6 dBm/ch to 50,295 per span for +4 dBm/ch) due to the maximum artificial FWM criterion. Again no significant dependence on the FFT length could be observed.

Figure 4.16 (left) shows the speedup factor for GPU-based simulations compared to CPU-based simulations. It can be observed from this figure that for all investigated scenarios the simulation time of the GPU-based implementation is lower than the CPU-based implementation (the speedup factor is always larger than 1). Furthermore, the speedup is increasing with a longer FFT-size. The reason for this is that a higher FFT length allows a better parallelization on the GPU architecture. We did not observe a significant (speedup) difference between different launch powers. For an FFT-size of 2^{18} a speedup of almost a factor of 50 has been measured with double precision accuracy. The simulation time on the CPU for such an FFT-size and a launch power of +4 dBm/ch is already 116 min for a single fiber span compared to 139 s with the GPU-based implementation.

In a second set of simulations we investigated simulations based on single precision (SP) arithmetic. SP allows an additional speedup (compared to DP) on current GPUs due to the fact that 8 SP processors are available for 1 DP unit. However, the CUDA^{TM} intrinsic FFT (CUFFT), exhibits a significantly higher error than CPU-based implementations. To increase the accuracy for SP we used our own FFT as shown before in Section A leading to a slightly lower performance for high FFT lengths than CUFFT as shown in Fig. 4.16 (left).

Results for SP simulations are depicted in Fig. 4.16 (right). Only FFT-lengths of 2^{15}–2^{18} are shown because SP outperforms DP only for higher FFT-lengths. The reason is that for shorter FFT lengths the fiber simulation time is dominated by copying all data from external memory to graphics memory. Our simulations show that the Q-factor deviation of our implementation remains below a value of 0.25 dB (equivalent to 2.9 % in non-logarithmic units) for the investigated scenarios. However, the error is significantly higher than the one shown in Fig. 4.15 (right) due to the use of SP. Compared to the intrinsic CUFFT implementation in single precision, though, our algorithm shows a much higher accuracy with the Q-factor deviation in the CUFFT implementation growing exponentially making it unusable for many simulation scenarios (especially with higher launch powers of more than 0 dBm/ch). If a Q-factor deviation of up to 0.25 dB can be tolerated in the simulations, our implementation provides a fast alternative with speedup of up to a factor of more than 200 compared to CPU-based simulations.

4.2.2 Stratified Monte-Carlo Sampling Technique

As shown in the last section a significant speedup compared to CPU-based calculations can be obtained for the SSFM by GPU parallelized implementation. The optimum speedup, however, requires utilizing SP accuracy simulations. Especially for a very high number of split-steps this leads to inaccuracies, which may exceed a certain desired limit. This is why in this section a stratified Monte-Carlo (MC) sampling technique is proposed making use of both SP and DP simulations and choosing the required amount of DP simulations automatically.

In the following, the theoretical background of the stratified sampling (SS) MC algorithm is given (based on [28]). For this purpose first the standard MC method is explained. Let I be a random variable with unknown average value $E[I]$ and unknown variance σ^2. Furthermore, $I = g(\vec{x})$ be a function (in this case the propagation of a signal through a fiber optical transmission system) of the random vector $\vec{x} = (x_1, \ldots, x_M)$ taking values from the input space Ω. Then the maximum-likelihood estimator of $E[I]$ is the mean value of the samples

$$\bar{I} = \frac{1}{n} \cdot \sum_{k=1}^{n} I^{(k)} \quad (4.2\text{-}9)$$

where $I^{(k)} = g(\vec{x}_k)$ with $k = 1, \ldots, n$. The vector $\vec{x}_k$ represents an independent realization of the random vector $\vec{x}$. This approach is also called MC estimation. $\bar{I}$ is an unbiased estimator of $E[I]$ with standard deviation:

$$\bar{\sigma} = \sqrt{\frac{\sigma^2}{n}}. \quad (4.2\text{-}10)$$

It is obvious that the standard deviation is decreasing with an increasing number of samples, which is one of the key features of an MC simulation.

As σ^2 is unknown, usually it is calculated with the help of the maximum likelihood estimator:

$$\sigma^2 \cong \frac{1}{n-1} \cdot \sum_{k=1}^{n} \left(I^{(k)} - \bar{I} \right)^2. \quad (4.2\text{-}11)$$

For a good estimate of the average value $\bar{I}$ the number of samples has to be high ($n \gg 1$), and the distribution of $\bar{I}$ resembles a normal distribution. This allows defining confidence intervals with the following probability density relation:

$$P((\bar{I} - \varepsilon \cdot \bar{\sigma}) \leq E[I] \leq (\bar{I} + \varepsilon \cdot \bar{\sigma})) = C \quad (4.2\text{-}12)$$

where C is the confidence level of the estimator. From the central limit theorem it can be derived that $\bar{I}$ will have approximately Gaussian statistics and the following relation holds:

$$\varepsilon = \sqrt{2}\mathrm{erfc}^{-1}(1 - C) \quad (4.2\text{-}13)$$

where erfc is the complimentary error function. A confidence level of e.g. 95 % results in $\varepsilon = 1.96$. Furthermore, a relative error at confidence level C can be defined as

$$\Delta_n = \frac{\varepsilon \bar{\sigma}}{\bar{I}}, \quad (4.2\text{-}14)$$

which is used as an abort criterion. In an MC simulation n is increased until $\Delta_n < \Delta$, where Δ is a target relative error. In the following we assume $\varepsilon = 1.96$ and $\Delta = 0.1$.

In our case the MC method is applied to an optical transmission system, and the random vector $\vec{x}$ contains all random variables involved in the simulation. This comprises for example the random bit sequences at the pulse generators as

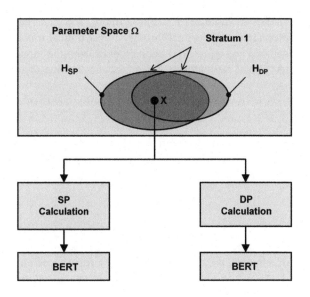

Fig. 4.17 Sketch of the different strata. Stratum 1 contains all sets in the input space H that lead to bit errors (adapted from [28])

well as the additive (numerical) noise realizations at the EDFAs. In this case, n is the total number of simulated bits, and I becomes an error indicator equal to 1, if a bit error occurs and 0 otherwise. The estimated average value $E[I]$ is equivalent to the bit error ratio. As in numerical simulations of fiber optical transmission systems using the SSFM usually bit blocks are processed, and not single bits are propagated, each of the bits contained in a block can be used to update the average value of I.

The stratified sampling (SS) approach is an improvement to the standard MC calculation. It partitions the sample space into r disjoint strata (compare Fig. 4.17). On each of these strata independent MC simulations are executed. The key assumption is that the variances in the individual strata are lower than in the entire sample space [28]. This allows a more efficient calculation compared to the standard MC approach.

For the SS approach some additional definitions are needed. Let p_s be the probability with which stratum s is visited, where $s = 1, \ldots, r$. Furthermore, $\bar{I}_s$ is the average value of I within stratum s obtained from an MC calculation. Analog to Eqs. (4.2-10) and (4.2-11) $\bar{\sigma}_s$ and σ_s^2 can be defined for each stratum. From the total probability law afterwards the global sample mean can be estimated as [28]:

$$\bar{I} = \sum_{s=1}^{r} p_s \cdot \bar{I}_s. \qquad (4.2\text{-}15)$$

The variance of such an estimator is given by

$$\bar{\sigma}^2 = \sum_{s=1}^{r} p_s^2 \cdot \bar{\sigma}_s^2. \qquad (4.2\text{-}16)$$

4.2 Parallelization of a Simulation on a Graphics Processing Unit

In our case the SP GPU accelerated simulation (fast system) is used to partition the entire sample space into strata. In the following only two strata are used. Stratum 1 contains all X_k for which $g(X_k) = 1$ (one or more bit errors during transmission), and stratum 2 is the complimentary set with all X_k leading to error free transmission.

The better the fast system approximates the true system (DP simulation), the better the overlap of the strata of both systems.

The probability p_s ($s = 1, 2$) of selecting each stratum is estimated from runs of the fast system with independent realizations of X_k, where $p_2 = 1-p_1$, and p_1 being the ratio of the number of visits of stratum 1 and the total number of fast simulation runs. For a subset of these X_k, MC simulations for the true system are started yielding $\bar{I}_s$ and $\bar{\sigma}_s^2$. With these estimations the global mean (Eq. 4.2-15) and the global variance (Eq. 4.2-16) are determined. The probability for calculating a certain realization X_k with the slow system (DP simulation) is given by $\alpha_1 = \min(1, \sigma_1/\sigma_2)$ and $\alpha_2 = \min(1, \sigma_2/\sigma_1)$. The stratum with the higher variance is preferred for simulations with the slow system in this way (more details can be found in [28]. If the fast and the true systems (i.e. SP and DP accuracy simulations) lead to very similar results, $\bar{I}_2 \approx 0$ and also the variance is $\sigma_2^2 \approx 0$. This means that X_k from stratum 1 are simulated with a high probability using the true system, whereas X_k from stratum 2 are skipped (for true system simulations). A flowchart of the SS MC approach is depicted in Fig. 4.18.

The SP simulation offers a significant speedup compared to DP simulations only for high FFT lengths $\geq 2^{15}$ (compare Fig. 4.16, left). The reason is that for shorter FFT lengths the simulation time is dominated by the transfer between main and graphics memories. The SS MC approach—using SP simulations as a fast system—should therefore only be performed in this operational region.

A typical setting for a simulation uses at least 32 samples / bit. Thus, the blocks of the sampled signal contain $\geq 2^{10}$ bits (if FFT lengths $\geq 2^{15}$ are desired). For a high BER ($\gg 10^{-3}$) consequently all blocks would be associated with stratum 1 because it is highly likely that they contain errors. For such a case, the SS MC approach would converge to a standard MC approach with the overhead of the fast system. Luckily, the desired BER (pre-FEC) is typically in the range of 10^{-3}–10^{-5} so these cases are very rare. Furthermore, GPU accelerated simulations allow simulating a transmission system with a high number of channels (leading to prohibitively large simulation times on purely CPU-based simulations) requiring more samples per bit and reducing the number of bits per block for the same FFT vector length.

In the following we use an exemplary 20 span transmission system with one 112 Gb/s coherent polmux-QPSK channel surrounded by 10 WDM 10.7 Gb/s NRZ-OOK channels on a 50 GHz grid [26]. As transmission fiber NZDSF has been assumed. To simulate the worst-case signal degradation all NRZ channels are co-polarized at the transmitter, and negligible PMD along the transmission line has been assumed. The dispersion compensation map has been optimized for 10 Gb/s transmission and −480 ps/nm pre-compensation has been used as well as 50 ps/nm/span undercompensation. At the receiver the accumulated dispersion has

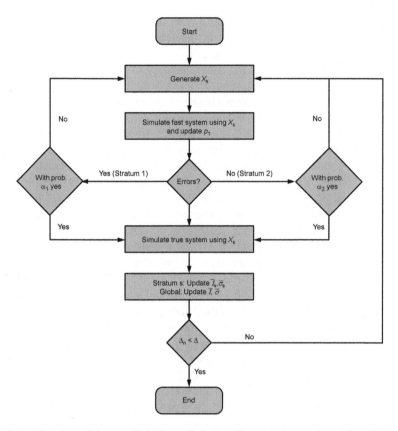

Fig. 4.18 Flowchart of the stratified Monte-Carlo sampling technique (adapted from [28])

been returned to 0. The BER has been assessed for the 112 Gb/s CP-QPSK center channel only, as this one is degraded worst. The launch power into the transmission fiber has been varied in our simulation from −6 dBm/ch to −1 dBm/ch leading to a total number of 2,300 split-steps for the former and 82,000 for the latter. The launch power into the DCF has always been selected to be 5 dB below the launch power into the NZDSF. A block length of 2,048 bit for each QPSK tributary has been used and 64 samples per bit (leading to an FFT-length of 2^{17}).

The results are depicted in Fig. 4.19. It can be observed from Fig. 4.19 (left) that the BER estimates obtained from MC simulations in double precision and the SS MC method are almost identical for all simulated launch powers. For pure single precision based simulations the deviation between the two curves would have increased with the launch power due to the higher number of split-steps required. In Fig. 4.19 (right) the speedup of the SS MC method is shown. It can be seen that a high speedup of up to 180 compared to CPU-based MC simulations can be obtained. For the chosen block length the speedup is increasing with a lower BER because in this case entire blocks of 2,048 bits are transmitted error free, which is advantageous for the SS approach distinguishing strata containing blocks

4.2 Parallelization of a Simulation on a Graphics Processing Unit

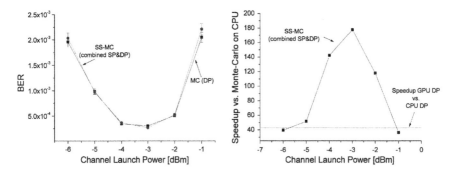

Fig. 4.19 BER vs. channel launch power (*left*) and speedup vs. channel launch power (*right*)

Table 4.2 Number of runs for the MC and SSMC simulations. A confidence level of the results of 95% ($\varepsilon = 1.9$) has been used

Channel launch power [dBm]	MC runs	SSMC DP runs	SSMC SP runs
−6	107	91	94
−5	186	119	133
−4	629	126	248
−3	858	145	240
−2	401	103	164
−1	69	64	64

with errors and error free ones. The exact numbers of MC or SSMC runs with 2,048 bits each are listed in Table 4.2.

To check the robustness of our approach we performed a simulation (for −3 dBm channel launch power) with an (deliberately selected) extremely high number of 880,000 split-steps by adjusting the maximum admissible step-size. It is expected that for such a high number of split-steps the deviation between SP and DP is high, and SP accuracy will not be sufficient (compare [27]). The combination of SS MC and our FFT algorithm, however, shows a BER of 2.417e-4 (SS MC on GPU) compared to 2.416e-4 (CPU MC in DP) at a speedup of 44.52 compared to the CPU making our approach suitable also for these high numbers of split-steps.

4.3 Analytical Modeling of the Signal Quality

The acceleration techniques presented in the last sections all apply to simulations based on numerical solving of the NLSE. Such an approach is very versatile as it does not make any assumptions on the transmitted pulses (e.g. bit rate or modulation format). If, however, the signal quality needs to be assessed on-the-fly—as it is required for online operation of a fiber optical network considering physical-layer impairments—the simulation time may still be too long. Also in the design phase a very fast assessment of the signal quality may be needed.

In this section another way of assessing the signal quality is presented based on analytical or semi-analytical models of the different degradation effects occurring during transmission along a fiber. Generally, it is not possible to solve the NLSE analytically for arbitrary input pulse shapes. For some cases, however, it is possible to find fast analytical models. The idea is to reduce the complexity of the NLSE by assessing the different degradation effects, e.g. chromatic dispersion, PMD, fiber nonlinearities (compare Eq. 3.1-32) separately neglecting dependencies between the different effects deliberately. In the following subsection analytical approaches for describing the Q-factor are presented including the signal degradation due to ASE-noise (accumulated) group-velocity dispersion, polarization mode dispersion, cross-phase modulation, four-wave mixing and stimulated Raman scattering induced crosstalk. As modulation format 10 Gb/s NRZ-OOK is assumed, which is still the most widely deployed modulation format in today's transmission systems.

The analytical models presented in the following subsections (based on our publications [29–32]) are also used in the next chapter to enable on-demand routing of wavelength granularity demands taking into account physical-layer impairments.

4.3.1 Linear Degradation Effects

If data is transmitted through a transparent optical network, degradation effects may accumulate over a large distance. The major linear impairments GVD, PMD and ASE-noise are presented in this section.

A. Group Velocity Dispersion

GVD leads to broadening of the pulses in time domain (compare also Sect. 3.1.1). This results in intersymbol interference (ISI) with the neighboring pulses and a decrease of the amplitude due to energy conservation. Because GVD is a deterministic effect an eye opening penalty (EOP) is used to estimate the degradation of the pulses. Usually DCFs are deployed to compensate for the accumulated dispersion. However, the dispersion slope of the DCF and the transmission fiber do not match perfectly in many cases leading to uncompensated chromatic dispersion. For RZ modulation with unchirped Gaussian-shaped pulses the EOP can be written as

$$EOP_{RZ} = \frac{\sigma_{deg}}{\sigma_{org}} = \sqrt{1 + \frac{(\beta_2 \cdot l)^2}{4\sigma_{org}^4}}. \qquad (4.3-1)$$

For the derivation of Eq. 4.3-1 the worst-case condition has been considered, which is given by a single '1', which is framed by two '1's. In (4.3-1) σ_{deg} and σ_{org} are the standard deviations of the degraded pulse after traveling a distance l with a GVD parameter β_2 and the original pulse at the transmitter, respectively.

For NRZ pulses with squared-cosine edges and a roll-off factor R the following equation can be derived heuristically

4.3 Analytical Modeling of the Signal Quality

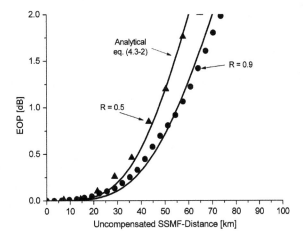

Fig. 4.20 EOP due to uncompensated GVD for a 10 Gb/s-NRZ system using SSMF. Symbols represent simulation results, the solid line the analytical results

$$EOP_{NRZ} = 1 + \left((\beta_2 \cdot l)^2 \left(B_R \cdot \sqrt{8 \ln 2}\right)^4\right)^{R+0.8} \quad (4.3\text{-}2)$$

where B_R stands for the bit rate. Equation 4.3-2 is a good approximation for EOP values smaller than 2 dB as can be seen from Fig. 4.20. Because GVD is a linear effect the accumulated dispersion mismatch can be added up along the entire path from the transmitter to the receiver.

B. Polarization Mode Dispersion

PMD occurs due to small disruptions of the symmetry of the fiber (compare Sect. 3.1.7). These can originate from the production process but also from changes of the ambient temperature or vibrations. PMD is a statistical effect. Due to PMD the two orthogonal modes exhibit different propagation constants. The linear degradation due to PMD can be assessed by an EOP. The statistics of PMD is described by the mean differential group delay (DGD). The DGD accumulates by a square law along the spans.

1. First-Order Polarization Mode Dispersion

In Ref. [33] the relationship between power penalty and DGD has been shown, which enables to calculate the PDF of the EOP.

$$EOP_{dB} = A \cdot \frac{\Delta\tau^2 \cdot \gamma \cdot (1-\gamma)}{T_{FWHM}^2} = A \cdot f(\Delta\tau, \gamma) \quad (4.3\text{-}3)$$

In Eq. 4.3-3 the DGD is given by $\Delta\tau$, γ defines the relative power launched in the two principle states and A is a pulse form factor, which is dependent on the pulse form and the receiver characteristics [34]. For a Maxwellian distributed DGD and a random power splitting, a negative exponential PDF can be defined for the EOP in dB [34]

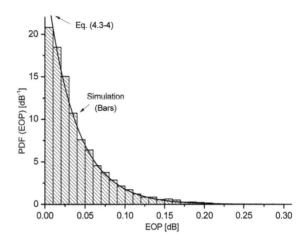

Fig. 4.21 PDF of the EOP from numerical simulations and the analytical results. Gaussian-shaped pulses, d = 0.4, T$_{FWHM}$ = 40 ps, A = 12

$$PDF(EOP_{dB}) = \eta \cdot \exp(-\eta \cdot EOP_{dB}) \quad (4.3\text{-}4)$$

with

$$\eta = \frac{16}{A \cdot \pi \cdot \langle\Delta\tau\rangle^2 \cdot B_R^2} \quad (4.3\text{-}5)$$

where $<\Delta\tau>$ is the mean DGD and B_R the bit rate.

The analytical PDF Eq. 4.3-4 was verified by Monte-Carlo simulations, and good agreement can be observed from Fig. 4.21.

2. *Second-Order Polarization Mode Dispersion*

Since first-order PMD can be compensated, it is interesting to know the behavior of pulse broadening due to second-order PMD (SOPMD). The signal degradation due to higher order PMD is given by the dependence on the frequency. For the case of fully-compensated first-order PMD a relation similar to Eq. 4.3-3 has been derived [35]:

$$EOP_{dB} = C|\Omega_\omega|^2 \quad (4.3\text{-}6)$$

Because the power is coupled to a PSP the coupling factor γ could be set to zero. The pulse form factor C can be determined by computer simulations [35]. Ω_ω represents the frequency deviation of the PMD vector and is referred to as SOPMD vector in the following. With Eq. 4.3-6 and the knowledge of the statistics of the SOPMD vector [36], the PDF of the EOP can be calculated by [35]:

$$\text{PDF}(EOP_{dB}) = \frac{16}{\langle\Delta\tau\rangle^4 \cdot \pi \cdot C} \operatorname{sech}\left(\frac{4}{\langle\Delta\tau\rangle^2}\sqrt{\frac{EOP_{dB}}{C}}\right) \cdot \tanh\left(\frac{4}{\langle\Delta\tau\rangle^2}\sqrt{\frac{EOP_{dB}}{C}}\right). \quad (4.3\text{-}7)$$

4.3 Analytical Modeling of the Signal Quality

In the following section it is shown how PMD can be included in the Q-factor calculation.

C. Amplified Spontaneous Emission-Noise

Erbium-doped fiber amplifiers (EDFA) always degrade the optical signal-to-noise ratio (OSNR). This comes from the fact that spontaneous emission inevitably occurs along stimulated emission. The OSNR can be calculated as already shown in Sect. 3.1.2, Eq. 3.1-18. To describe the degradation of the OSNR a noise figure F_o is typically used, which is defined as the ratio of the OSNR before and after the amplifier.

If a Gaussian distribution of the noise is assumed, the BER can be calculated by Eq. 3.1-55. If the ASE-channel beat noise is assumed to be dominant, a relationship between OSNR, EOP and Q-factor can be defined [37].

$$Q = \sqrt{\frac{OSNR}{EOP} \cdot \frac{\Delta f_{opt}}{\Delta f_{el}} \cdot EXTP} \qquad (4.3\text{-}8)$$

where *EXTP* is the extinction ratio penalty.

In this Q-factor definition polarization effects have been neglected. To take into account the additional impairments due to first-order PMD in [34] extensions to Eq. 4.3-8 are given:

$$Q = \frac{1}{EOP_{PMD}} Q_{w/oPMD}. \qquad (4.3\text{-}9)$$

Equation 4.3-9 assumes that PMD further degrades the inner eye-opening neglecting the changes in the noise variance. $Q_{w/oPMD}$ is the Q-factor as defined in Eq. 4.3-8 and EOP_{PMD} is given by Eq. 4.3-3, but now in linear units. With Eq. 4.3-4 a PDF of the Q-factor can be calculated [34]

$$\text{PDF}(Q) = \frac{10 \cdot \eta}{Q \cdot \ln 10} \cdot \left(\frac{Q_{w/o\_PMD}}{Q}\right)^{\frac{-10 \cdot \eta}{\ln 10}} \qquad (4.3\text{-}10)$$

If a minimal Q-factor is given, which is needed for error-free transmission, an outage probability OP can be calculated by [34]

$$\text{OP} = \int_0^{Q_{\lim it}} \text{PDF}(Q) dQ = \left(\frac{Q_{\lim it}}{Q_{w/o\_PMD}}\right)^{\frac{10 \cdot \eta}{\ln 10}} \qquad (4.3\text{-}11)$$

If SOPMD is taken into account instead of first-order PMD, a similar procedure has been described in [35].

4.3.2 Nonlinear-Degradation Effects

Higher input power into the transmission fiber allows mitigating noise-induced impairments. However, increasing the input power also shifts nonlinear impairments into the focus. The dominant nonlinear effects in 10 Gb/s NRZ systems are XPM and FWM. Additionally, stimulated Raman scattering (SRS) may induce a spectral tilt (and also crosstalk). All three effects are covered in this section. In this book only the most important formulas are given. The reader is referred to [6] for more details.

A. Cross-Phase Modulation

XPM is one of the main obstacles to reach high transparent transmission distances in WDM systems containing 10 Gb/s NRZ-OOK channels operating on SSMF. It causes a modulation of the optical phase of a given channel through the refractive index variation of the fiber when other channels are intensity modulated (see also Sect. 3.1.4). The modulation of the optical phase is converted into intensity modulation of the signal by dispersion [38], leading to a noise-like distortion. In the following an analytical model is presented, which is based on the considerations in [39], where XPM degradations are assessed by a transfer function in the frequency domain.

The analytical model in [39] can be simplified significantly for the case of negligible dispersion slope ($D_{probe} \approx D_{pump}$) and a large attenuation and fiber length product $\alpha \cdot L \gg 1$. These simplifications also hold for realistic system setups ($S \neq 0$) as will be shown below. In the following the index i represents the probe channel number and the index k for the pump channel number. The number of fiber spans is denoted by N. For a single pump channel with power $\tilde{P}_k$, the average power of the probe channel $\bar{P}_i$ and the power fluctuation introduced by XPM $\tilde{P}_{XPM}$ one arrives at a small-signal transfer function H_{XPM} defined as

$$H_{XPM,ik}(\omega) = \left| \tilde{P}_{XPM,ik} / \left(\bar{P}_i(0) \cdot \tilde{P}_k(\omega) \right) \right|. \quad (4.3\text{-}12)$$

The transfer function can be calculated as follows:

$$H_{XPM,ik}(\omega) = 2g_{net} \cdot \sum_{l=1}^{N} \gamma^{(l)} \exp\left(j\omega D_I^{(l-1)} \Delta \lambda_{ik}\right) \prod_{n=1}^{l-1} \left(\exp(-\alpha_l L_l) g_k\right)$$
$$\cdot \left(\frac{1}{(a_{ik})^2 + (2b_i)^2} [a_{ik}(C_i - 2D_i^{(l)}) - 2b_i)] + \frac{\sin(C_i)}{a_{ik}} \right) \quad (4.3\text{-}13)$$

with g_{net} being the net power gain from the transmitter to the receiver, $\gamma^{(l)}$ being the nonlinearity coefficient of the lth-fiber, $D_I^{(l-1)}$ the accumulated inline dispersion in front of the lth-segment, $\Delta \lambda_{ik}$ the channel spacing between channels i and k, L_l the length of the lth-transmission fiber, α_l the attenuation constant of the fiber, g_k the gain of the booster in front of the lth-transmission fiber and

$$C_i = \frac{\omega^2 \lambda_i^2}{4\pi c} D_R \quad (4.3\text{-}14)$$

$$D_i^{(l)} = \frac{\omega^2 \lambda_i^2}{4\pi c} D_I^{(l-1)} \quad (4.3\text{-}15)$$

4.3 Analytical Modeling of the Signal Quality

$$a_{ik} = \alpha - j\omega D \cdot \Delta\lambda_{ik} \quad (4.3\text{-}16)$$

$$b_i = \omega^2 D \lambda_i^2 / (4\pi c_0) \quad (4.3\text{-}17)$$

with λ_i being the wavelength of the ith channel, c_0 the speed of light, D_R the residual dispersion at the end of the system and D the dispersion parameter of the lth-fiber. Equation 4.3-13 has the great advantage that it allows to model systems with arbitrary dispersion compensation—varying on a span-by-span basis—and supports various fiber types in different links. The nonlinearity of the dispersion compensation fiber (DCF) has been neglected in all formulas. That is why N denotes the total number of spans and not the number of segments in contrast to the notation originally used in [39]. It is shown later by comparing the results from the original model [39] with the simplified version Eq. 4.3-13 that neglecting the nonlinearity of the DCF does not cause a considerable error. Furthermore, approximations have been used for the sine and cosine terms with $C_i - 2D_i$ because the argument is small, if a suitable pre-compensation is used. Equation 4.3-13 has to be evaluated for each pump channel $k = 1, .., M$. To obtain an analytical expression for the XPM-induced noise-like variance the following approach is used

$$\sigma_{XPM,i}^2 = \bar{P}(0)^2 \sum_{j=1, j \neq i}^{M} \frac{1}{2\pi} \int_{-\infty}^{+\infty} |H_{XPM,ij}(\omega)|^2 \cdot |H_{opt,filter}(\omega)|^2 \cdot PSD_j(\omega) d\omega \quad (4.3\text{-}18)$$

with the average power $\bar{P}(0)$ at the beginning of the fiber, the XPM transfer function $H_{XPM,ij}$, the filter function of the optical filter $H_{opt,filter}$, the total number of channels M and the power spectral density of the pump channel PSD_j [40]:

$$PSD_{NRZ}(\omega) = \frac{P_{max}^2}{4} \left[\left| \frac{\cos(R \cdot T\omega/2)}{1 - (R \cdot T\omega/\pi)^2} \frac{\sin(T\omega/2)}{T\omega/2} \right|^2 + \delta(\omega) \right] \quad (4.3\text{-}19)$$

with P_{max} being the peak power at the beginning of the fiber, R the cosine roll-off factor and T the bit duration.

In the following diagrams the values obtained from the analytical formulas are compared to the values obtained from solving the coupled nonlinear Schrödinger equation with a split-step Fourier (SSF) approach as shown in the previous sections. During the simulations all nonlinear effects except for the XPM degradation have been turned off. To achieve maximal decorrelation between neighboring channels the bit pattern is shifted by the ratio of the PRBS-length (here: 256 bits) and the total channel number (here: 9 channels) yielding a shift of 28 bits. As can be seen from Fig. 4.22 the analytical models, the "Cartaxo model" [39] and the simplified model from Eq. 4.3-13, both give a good estimation for the Q-factor. Also for other setups with higher input powers good agreement has been observed [6]. Please note that only the Q-factor due to XPM impairments is given. ASE-noise has been neglected deliberately at this point. Especially important is the fact that the Q-factor is calculated with a relatively small error in the maximum. This is the point, which is most

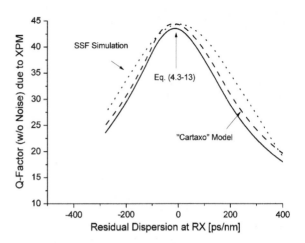

Fig. 4.22 Comparison of the Q-factors (w/o noise) for a 7-span NRZ 10 Gb/s transmission system. The Q-factor due to XPM is plotted for the center channel

desirable because the system performance is optimal. If the limits set by the computational time are strict and the XPM-induced variance has to be computed on-the-fly, further simplifications have to be introduced.

We have shown that it is possible to obtain a scaling law depending on the number of WDM channels. The idea is to calculate a two channel system with only one pump channel and to scale the derived results for an arbitrary number of pump channels with the help of the weighting factor W_M (more details can be found in [6, 41]). With the help of this approach the computational time can be further reduced significantly because the number of summations over different channels in Eq. 4.3-13 is only two. This makes the approach especially suitable for on-the-fly routing purposes.

The analytical XPM model described above has also been experimentally verified (more details can be found in [29]). For this purpose a recirculating loop setup has been used. Eight NRZ-OOK modulated 9.953 Gb/s channels have been spaced in a 50 GHz grid with an additional unmodulated channel in the center (pump-probe setup). All channels had parallel polarization at the loop input. The channels were decorrelated by a pre-DCF. The variances as well as the mean values of the center channel power level have been measured at different span counts and post-compensation values. The amplitude distortion, measured by the variance of the cw-channel power, has been taken as a figure of merit for the signal degradation due to XPM. In the experiment 7 SSMF spans have been cascaded with a pre-DCF of –170 ps/nm. Amplifiers were located before each transmission fiber and each DCF. In each span an undercompensation of 50 ps/nm has been employed. The booster amplifiers had an output power of 3 dBm/ch for all transmission fibers and 0 dBm/ch for the DCFs. The span length has been chosen to be 70 km. In the following, only the values for the center channel will be discussed, which were displayed on a sampling oscilloscope. This channel is also most affected by XPM and thus provides a worst-case estimation.

In Fig. 4.23 the values obtained from the simplified analytical formula Eq. 4.3-13 and the loop experiment are depicted. Also shown are the values

4.3 Analytical Modeling of the Signal Quality

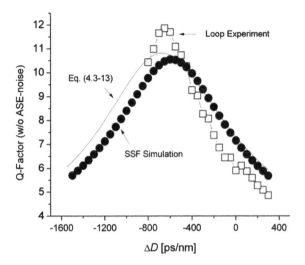

Fig. 4.23 Comparison of the Q-factors (w/o noise) for a 7-span NRZ 10 Gb/s transmission system. The Q-factor due to XPM is plotted for the center channel

obtained from a simulation employing the split-step Fourier (SSF) method. In the simulations only the nonlinear degradations due to XPM have been considered (separated channels approach). The depicted Q-factor was calculated as the ratio of the mean value and the standard deviation of the probe channel power level. The results of the loop-experiment are depicted only in the range of $\Delta D = +300..$ -800 ps/nm due to the range of the tunable dispersion compensator, which was incorporated after the loop. ΔD specifies the added amount of post-dispersion compensation. From Fig. 4.23 we conclude that the analytical expression yields a good estimation for the Q-factor. The optimal post-compensation of the dispersion is predicted correctly, and there is only a slight difference in the height of the maximum, which is tolerable for a fast analytical assessment.

B. Four-Wave Mixing

Four-wave mixing (FWM) is one of the dominant degradation effects in WDM systems with dense channel spacing and low local chromatic dispersion. If in a WDM system the channels are equally spaced in frequency, the new waves generated by FWM will fall at channel frequencies and thus will contribute to crosstalk. In case of full inline dispersion compensation the FWM crosstalk becomes maximum since the FWM products add coherently in each span.

The analytical model presented for the assessment of the FWM degradation in this book is based on a single span cw approximation [37]. The amplitude of an FWM mixing product at frequency m, which is generated by waves i, j and k, can be calculated in the general case from the following equation [37]:

$$A_{m,ijk}(l_1, l_2) = j\gamma \left(\frac{d}{3}\right) A_i(0) A_j(0) A_k^*(0) \cdot \int_{z=l_1}^{l_2} \exp(-\alpha z - j\Delta\beta_{ijk} z) dz \quad (4.3\text{-}20)$$

Here A is the amplitude of the envelope of the different waves, γ is the nonlinearity constant of the fiber and α the attenuation constant. l_1 and l_2 are the absolute positions on the fiber, measured from the beginning of the transmission system. The degeneracy factor d has a value of 3 for $i = j$ (two-tone product) and a value of 6 for $i \neq j$ (three-tone product). In Eq. 4.3-20 the amplitude distortion due to dispersion has been neglected. This is valid for low-dispersion fibers such as DSF or NZDSF and moderate bit rates (i.e. 10 Gb/s systems) where FWM degradation becomes dominant. The phase matching between the different waves can be calculated from [37]:

$$\Delta\beta = \beta_i + \beta_j - \beta_k - \beta_{ijk} = \frac{2\pi\lambda^2}{c_0}(f_i - f_k)(f_j - f_k) \times \left[D - \frac{\lambda^2}{c_0}\left(\frac{f_i + f_j}{2} - f\right)S\right] \quad (4.3\text{-}21)$$

To take into account the random nature of the modulated signal three tone products are weighted by a factor of 1/8 and two tone products by a factor of 1/4. This comes from the probability in a PRBS, which is 1/2 for a logical '1'.

The power of a single FWM product at the end of a single piece of fiber of length L can be calculated by:

$$P_{m,ijk}(L) = \left(\frac{d}{3}\gamma\right)^2 e^{-\alpha L} P_i(0) P_j(0) P_k(0) \left|\frac{\exp(-\alpha L - j\Delta\beta_{ijk}L) - 1}{-\alpha - j\Delta\beta_{ijk}}\right|^2. \quad (4.3\text{-}22)$$

An extension to multi-span systems has been shown in [42]. From the sum of all FWM products falling in the regarded channel a variance can be computed, which can be treated similar to ASE-noise.

$$\sigma^2_{FWM} = 2P_{ch} \sum_{ijk} P_{m,ijk} \quad (4.3\text{-}23)$$

From Fig. 4.24 it can be seen that the degradation due to FWM mainly causes a beating on the '1'-level, which can be approximated by a Gaussian-shaped PDF. This also underlines the treatment of FWM as an additional noise variance.

To verify the results of the analytical model another experiment has been carried out. Again a recirculating loop setup has been used. Twenty NRZ-modulated 9.953 Gb/s channels have been spaced on a 50 GHz grid (1531–1538 nm). All channels had a parallel polarization at the loop input, which represents the worst-case situation. In the experiment a pre-DCF has been employed to decorrelate the channels. Amplifiers were located before each transmission fiber and each DCF. Inside the loop three spans of Lucent TrueWave® Classic were cascaded with an average dispersion of 2.4 ps/(nm km). Full-inline compensation (FOCS) has been used (0 ps/nm in the first span, –6 ps/nm in the second span and 0 ps/nm in the third span). No additional post-compensation was employed at

4.3 Analytical Modeling of the Signal Quality

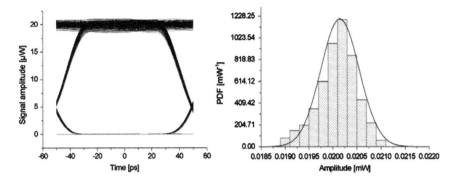

Fig. 4.24 Eye diagram and histogram after 3 spans, 21 channels, 3 dBm/ch, TrueWave Reduced Slope fiber, 50 GHz channel spacing, $D_{res} = -20$ ps/nm/span

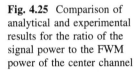

Fig. 4.25 Comparison of analytical and experimental results for the ratio of the signal power to the FWM power of the center channel

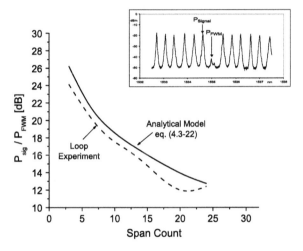

the output of the loop because this does not affect the measured FWM power. The booster amplifiers had an output power of 1 dBm/ch for all transmission fibers and −2 dBm/ch for the dispersion compensating fibers (DCF). The lengths of the three spans were 68, 68 km and 64 km, respectively.

In the experiment the FWM power was measured outside of the signal spectrum with an optical spectrum analyzer. For this purpose the center channel was omitted so that at this position solely the FWM power could be measured (inset in Fig. 4.25). The FWM power is considered as a figure of merit for the signal degradation because it will lead to crosstalk whenever it falls directly onto the signal channel. The discrepancy between the analytical curve and the measured one (Fig. 4.25) stems mainly from the fact that the DCFs did not match the exact dispersion compensation map, but varied slightly. Also the DCF nonlinearity was neglected in the analytical model. For a more detailed discussion of the experiment the reader is referred to [30].

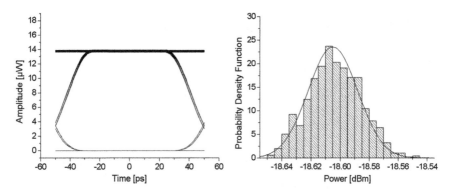

Fig. 4.26 Eye diagram and histogram of the highest frequency channel after 1 span. 64 channels, −1 dBm/ch, D = 1.4 ps/(nm. km), 200 GHz channel spacing

C. Stimulated Raman Scattering

In WDM systems with a high channel count and a large spectral bandwidth SRS leads to a tilt in the spectrum, and the random bit sequences on the different WDM channels cause crosstalk [43].

The tilt of the spectrum can be calculated analytically by the following equation

$$a_{SRS,tilt,dB} = 10 \cdot \log e \cdot P_{ch} \frac{l_{eff}}{K \cdot A_{eff}} \frac{\partial g_R}{\partial f} \sum_{i=2}^{M} (f_i - f_1). \quad (4.3\text{-}24)$$

In Eq. 4.3-24 the Raman gain factor has been approximated by a triangular shape [44]. P_{ch} stands for the channel input power, l_{eff} is the effective fiber length, A_{eff} the effective area, K the polarization factor, g_R the Raman gain factor and f_i the channel frequency.

Figure 4.26 shows that the SRS crosstalk (split-up of the eye-lines) is negligible on dispersive fiber outside the dispersion zero. The PDF of the '1's can be approximated by a Gaussian distribution to the first order. The spectral tilt can be overcome by gain-flattening filters in the amplifiers. That is why we believe that in constraint-based routing SRS can be omitted in most cases, if transient effects occurring when wavelength channels are switched on and off, are well controlled.

4.3.3 System Example

The nonlinear fiber effects, described in the previous section, can be included easily into the Q-factor calculation based on Eqs. 4.3-8 and 4.3-9. The analytical models for XPM and FWM offer the possibility to calculate a variance of the eye lines. The additional variance due to XPM can be estimated by Eq. 4.3-18 and the variance due to FWM by Eq. 4.3-23. These variances can be added to the variance stemming from ASE-noise.

4.3 Analytical Modeling of the Signal Quality

Table 4.3 Reference system parameters

	System 1	System 2
P_{ch} [dBm]	0	3
α [dB/km]	0.2	0.2
D [ps/nm/km]	17 (SSMF)	4 (NZDSF)
S [ps/nm²/km]	0.056	0.045
l [km]	100	100
D_{DCF} [ps/nm/km]	−102	−102
S_{DCF} [ps/nm²/km]	−0.23	−0.23
l_{DCF} [km]	16.667 (0 ps/nm/span)	3.43 (+50 ps/nm/span)

To further investigate the capability of the fast analytical models in this section results from full numerical simulations are compared to the ones from the analytical models. As a reference system a nine channel 10 Gb/s NRZ OOK system has been chosen. Two setups have been selected exemplarily. The first setup is based on SSMF with a relatively low channel input power (0 dBm/ch). The second setup operates on NZDSF with a higher input power (3 dBm/ch) deliberately fortifying the nonlinear effects. In the latter case a distributed undercompensation scheme with 50 ps/nm/span has been deployed (Table 4.3).

We assumed all EDFAs to have a noise figure of 4.5 dB. A precompensation of −170 ps/nm has been used. At the receiver side a 25 GHz first-order Gaussian-shaped optical demultiplex filter has been used and a 7 GHz electrical Bessel filter of 10th order after the PIN diode. The residual dispersion has not been optimized at the receiver ($D_{res} = 0$). Always the center channel has been analyzed.

In the simulations the SSFM has been used to solve the nonlinear Schrödinger equation. A total number of 256 bits has been launched at the transmitter. Polarization effects have been neglected in the simulations. At the receiver side a bit error ratio tester has been used to compute the Q-factor. The results have been compared to the fully analytical Q-factor calculated from the models presented in the previous sections taking into account ASE-channel beat noise, XPM and FWM. No efforts have been made to optimize the residual dispersion at the receiver. This also reflects the situation in ASONs, where different channels may have different origins and different accumulated dispersion values. Optimizing the dispersion at the receiver could thus only be achieved on a per-channel basis, which is cost intensive (if not digital signal processing is used anyway).

Fig. 4.27 shows that the analytical model overestimates the Q-factor in the low degradation regime (Q > 9), corresponding to very low bit error ratios. However, both curves agree very well for lower Q-factors. In real system typically a minimal (linear) Q-factor of 3.5–4.5 (before FEC) may be required. Both, the analytical model and the simulation, show that 19 spans can be bridged for this case.

To move to the limits of our analytical models a strongly nonlinearly limited regime has been chosen deliberately in the second setup. This time NZDSF has been used, and the input power has been increased to 3 dBm/ch. For dispersion compensation a distributed undercompensation scheme (−50 ps/nm/span) has been

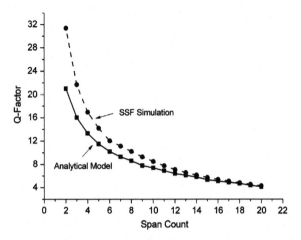

Fig. 4.27 Comparison of Q-factors from simulation and analytical model. SSMF, 9 ch, 0 dBm/ch, 50 GHz channel spacing, full-inline dispersion compensation (System 1)

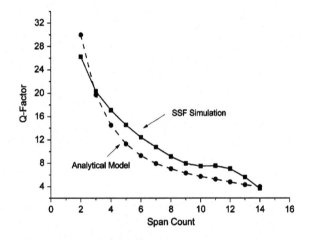

Fig. 4.28 Comparison of Q-factors from simulation and analytical model. NZDSF, 9 ch, 3 dBm/ch, 50 GHz channel spacing, −50 ps/nm/span (System 2)

assumed. As can be seen from Fig. 4.28 this time the analytical models underestimate the Q-factor in a large region. This stems from an overestimation of the XPM impairments in the analytical model for this setup. Similar behavior can also be observed from Fig. 4.22. Outside of the optimal residual dispersion the analytical XPM model tends to overestimate the impairments. However, in Fig. 4.28 a maximal distance of 13 spans has been predicted correctly (for Q > 4.5). In the simulated curve a slight improvement in the Q-factor can be seen for 11 and 12 spans. The analytical FWM model shows a similar behavior, which can be ascribed to the distributed undercompensation scheme.

Both examples show that analytical models—taking into account ASE-noise and the isolated effects of XPM and FWM only—can be used to give a good estimate for the signal quality.

4.4 Summary and Discussion

The complexity of optical transmission systems is increasing steadily. Numerical simulations for optimizing the design of such systems are thus indispensable. In this chapter several methods for reducing the simulation time for the assessment of the signal quality in a fiber optical transmission system have been presented enabling a more efficient network design. The optimization (i.e. reduction) of the simulation time is highly desirable with current numerical system simulations typically lasting for several days or weeks.

The chapter started with the presentation of a meta-heuristic based optimization algorithm. It allows searching a large parameter space efficiently by selecting sampling points sparsely, interpolating the results and using an optimization algorithm on this function. The accuracy of the interpolation function is improved iteratively. It has been shown that this meta-heuristic approach can reduce the number of numerical simulations by a factor of more than 100 for typical transmission scenarios.

In the following sections the parallelization of simulations on graphics processing units has been investigated. Our results proved that by a novel GPU-based implementation of the FFT with pre-computed twiddle factors, single precision simulations with the split-step Fourier method (SSFM) exhibit a high enough accuracy in most cases. These simulations offer a speedup of up to more than 200 compared to simulations on CPUs. Furthermore, a stratified Monte Carlo sampling approach has been presented. It can automatically choose the amount of required single and double precision accuracy simulations for a predetermined maximum relative error. With this approach also transmission system simulations with a very high number of split-steps can be performed with a high accuracy at the same time showing a significant speedup of up to more than 180 compared to CPU based simulations. Both techniques—GPU parallelization and stratified Monte Carlo sampling—have already been included in the commercially distributed simulation system for optical transmission systems *PHOTOSS* and are used intensively by industry partners.

In the last section of this chapter an alternative to numerical system simulations has been presented. By analytical models the signal quality after transmission may be estimated in the matter of seconds. The use of such approximation models is mandatory, if the performance of the system needs to be assessed on-the-fly. This is needed e.g. during the operation of a transmission system when the optical system performance shall be included into the routing decision as will be shown in the next chapter. The accuracy of the proposed models has been compared to laboratory experiments yielding a good agreement.

Increasing the computational power by the use of grid computing and of a high number of processing cores is widely available today. The algorithms presented in this chapter are suited to both techniques. The meta-heuristic optimization can be parallelized on individual PCs connected in a computing grid, whereas the GPU-based SSFM implementation inherently makes use of a high number of processing cores. The combination of both methods can join the individual speedups of both methods leading to a gain of up to more than a factor of 20,000 (comparing GPU-based simulations with the meta-heuristic optimization algorithm to grid-search simulations on a CPU).

It shall also be mentioned that the meta-heuristic algorithm is very versatile and can be utilized for various optimization problems also of entirely different nature than optical transmission systems. This can be achieved by simply exchanging the numerical simulation program by another simulation engine.

In which direction the development of GPUs will head will be interesting to follow. The CEO of NVIDIA Jen-Hsun Huang has recently announced an eight times increase in GFlops/Watt during the next 3 years in his keynote speech on the roadmap of future graphics cards [45], while keeping the power consumption constant (at approximately 250–300 W). The upcoming generations of GPUs are named Kepler and Maxwell already indicating that scientific computing will be a major target group for these cards. As the number of graphics cards used in computer gaming, however, is certainly much larger than in scientific computing the future will tell whether the (costly) development of GPUs for general purpose computing is economically viable. Independent of the direction of the future development of graphics cards the parallel SSFM implementation presented in this chapter can be used on gaming oriented GPUs (with a much higher number of single precission accuracy processing cores) and also on GPUs aimed at scientific computing (with increased double precission performance).

The analytical models presented in this chapter have been developed for the 10 Gb/s NRZ-OOK modulation format. Linear transmission impairments such as accumulated chromatic dispersion, polarization mode dispersion and filter crosstalk may be modeled also for higher bitrates by (semi–) analytical approaches. For 40 Gb/s and 100 Gb/s transmission systems with novel modulation formats the analytical models for the nonlinear fiber effects, however, will have to be adapted. In the research community different approaches are discussed. One direction is to base the signal quality assessment on a very simple criterion such as the weighted nonlinear phase shift to capture the impact of the accumulated nonlinearities [46]. This approach is unbeatably fast, however, universal application to arbitrary heterogeneous transmission systems is questionable. Furthermore, the weighted nonlinear phase shift has to be adapted to novel modulation formats and bit rates by time consuming simulations or laboratory experiments. Especially modeling of hybrid transmission systems with a mix of 10 Gb/s NRZ-OOK legacy channels and 40 Gb/s or 100 Gb/s channels is limited with this procedure.

An alternative approach is to implement a mix of different analytical and fast numerical models to account for the individual transmission impairments and to combine these results to a single figure of merit afterwards. This will be shown in Sect. 5.4 for estimating the SPM-induced penalties of 40 Gb/s and 100 Gb/s transmission formats for use in physical-layer impairment based routing. Such a procedure is very flexible regarding heterogeneous dispersion maps, mixed fiber types, etc. In combination with a sophisticated numerical transmission simulator (e.g. using parallelization of the numerical calculation) modeling of the

SPM-induced penalties in less than a second for a typical transmission distance can be expected.

For hybrid transmission systems with a mixture of 10 Gb/s NRZ-OOK and higher bit rate phase modulated channels accurate modeling of XPM-induced penalties is crucial. The extension of the analytical model for XPM may be a promising solution for this purpose (as proposed in [47]). Moreover, PolXPM crosstalk may become a limiting effect on polarization-multiplexed transmission of 112 Gb/s channels. Analytical models for PolXPM are subject of current research [48].

References

1. Pachnicke, S., Luck, N., Müller, H., Krummrich, P.: Multidimensional meta-model based optimization of optical transmission systems. IEEE J. Lightw. Technol **27**(13), 2307–2314 (2009)
2. Gaete, O., Coelho, L.D., Spinnler, B., Schmidt, E.-D., Hanik, N.: Global Optimization of Optical Communication Systems. European Conference on Optical Communication (ECOC), Brussels, Belgium (2008)
3. Alicherry, M., Nagesh, H., Poosala, V.: Constraint-based design of optical transmission systems. IEEE J Lightw Technol **21**(11), 2499–2510 (2003)
4. Frignac, Y., Antona, J.-C., Bigo, S., Hamaide, J.-P.: Numerical optimization of pre- and in-line dispersion compensation in dispersion-managed systems at 40 Gbit/s. Optical Fiber Communication Conference (OFC), ThFF5, Anaheim (2002)
5. Killey, R.I., Thiele, H.J., Mikhailov, V., Bayvel, P.: Reduction of intrachannel nonlinear distortion in 40-Gb/s-based WDM transmission over standard fiber. IEEE Photon Technol. Lett. **12**(12), 1624–1626 (2000)
6. Pachnicke, S., Reichert, J., Spälter, S., Voges, E.: Fast analytical assessment of the signal quality in transparent optical networks. IEEE J. Lightw. Technol. **24**(2), 815–824 (2006)
7. Louchet, H., Hodzic, A., Petermann, K., Robinson, A., Epworth, R.: Analytical model for the design of multispan DWDM transmission systems. IEEE Photon Technol. Lett. **17**(1), 247–249 (2005)
8. Frignac, Y., Antona, J.-C., Bigo, S.: Enhanced analytical engineering rule for fast optimization dispersion maps in 40 Gbit/s-based transmission. In: Proceedings of Optical Fiber Communication Conference (OFC), TuN3, Los Angeles, Feb 2004
9. Vorbeck, S., Schneiders, M.: Cumulative nonlinear phase shift as engineering rule for performance estimation in 160-Gb/s transmission systems. IEEE Photon Technol. Lett. **16**(11), 2571–2573 (2004)
10. Windmann, M., Pachnicke, S., Voges, E.: PHOTOSS: The Simulation Tool for Optical Transmission Systems, pp. 51–60. SPIE ITCOM, Orlando (2003)
11. Bigo, S.: Modelling of WDM Terrestrial and Submarine Links for the Design of WDM Networks. Optical Fiber Communication Conference (OFC), Anaheim (2006)
12. Aarts, E., Lenstra, J.K.: Local Search in Combinatorial Optimization. Wiley, Chichester (1997)
13. Kirkpatrick, S., Gelatt, C.D., Vecchi, M.P.: Optimization by simulated annealing. Science **4598**(220), 671–680 (1983)
14. McKay, M.D., Conover, W.J., Beckman, R.J.: A comparison of three methods for selecting values of input variables in the analysis of output from a computer code. Technometrics **21**, 239–245 (1979)
15. Morris, M.D., Mitchell, T.J.: Exploratory designs for computational experiments. J. Statist. Plann. Inference **42**, 381–402 (1992)

16. Shepard, D: A two-dimensional interpolation function for irregularly-spaced data, 23rd ACM National Conference, pp. 517–524 (1968)
17. Hardy, R.L.: Multiquadric equations of topography and other irregular surfaces. J. Geophys. Res. **76**(8), 1905–1915 (1971)
18. Amidror, I.: Scattered data interpolation methods for electronic imaging systems: a survey. J. Electron. Imaging **11**(2), 157–176 (2002)
19. Malach, M., Bunge, C.-A., Petermann, K.: Design rule for XPM-suppression in 10-Gb/s NRZ-modulated WDM transmission systems. IEEE LEOS Annual Meeting, pp. 481–482, Oct 2005
20. Kanter, D.: NVIDIA's GT200: Inside a Parallel Processor. http://ww.realworldtech.com/page.cfm?ArticleID=RWT090808195242, 2008
21. NVIDIA. NVIDIA GeForce GTX 200 GPU Architectural Overview, http://www.nvidia.com/docs/IO/55506/GeForce_GTX_200_GPU_Technical_Brief. pdf, 2008
22. Cooley, J.W., Tukey, J.W.: An algorithm for the machine calculation of complex fourier series. J. Math. Comput. **19**(90), 297–301 (1965)
23. Govindaraju, N. K., Lloyd, B., Dotsenko, Y., Smith, B., Manferdelli, J.: High Performance Discrete Fourier Transforms on Graphics Processors. In: ACM/IEEE Conference on Supercomputing, pp. 1–12, Nov 2008
24. Schatzmann, J.C.: Accuracy of the discrete Fourier transform and the fast Fourier transform. SIAM J. Scient. Comput. **17**(5), 1150–1166 (1996)
25. Hellerbrand, S., Hanik, N.: Fast Implementation of the Split-Step Fourier Method Using a Graphics Processing Unit. Optical Fiber Communication Conference (OFC), OTuD7, Mar 2010
26. Pachnicke, S., Chachaj, A., Remmersmann, C., Krummrich, P.: Fast Parallelized Simulation of 112 Gb/s CP-QPSK Transmission Systems using Stratified Monte-Carlo Sampling, accepted for Optical Fiber Communications Conference (OFC 2011), Los Angeles, March 2011
27. Pachnicke, S., Chachaj, A., Helf, M., Krummrich, P.: Fast Parallel Simulation of Fiber Optical Communication Systems Accelerated by a Graphics Processing Unit. IEEE International Conference on Transparent Optical Networks (ICTON 2010), Munich, Germany (2010)
28. Serena, P., Rossi, N., Bertolini, M., Bononi, A.: Stratified sampling Monte Carlo algorithm for efficient BER estimation in Long-Haul optical transmission systems. IEEE/OSA J. Lightw. Technol. **27**(13), 2404–2411 (2009)
29. Pachnicke, S., Hecker-Denschlag, N., Spälter, S., Reichert, J., Voges, E.: Experimental verification of fast analytical models for XPM-impaired mixed-fiber transparent optical networks. IEEE Photon Technol. Lett. **16**(5), 1400–1402 (2004)
30. Pachnicke, S., De Man, E., Spälter, S., Voges, E.: Impact of the inline dispersion compensation map on four wave mixing (FWM) - impaired optical networks. IEEE Photon Technol. Lett. **17**(1), 235–237 (2005)
31. Pachnicke, S., Gravemann, T., Windmann, M., Voges, E.: Physically constrained routing in 10-Gb/s DWDM networks including fiber nonlinearities and polarization effects. IEEE J. Lightw. Technol. **24**(9), 3418–3426 (2006)
32. Pachnicke, S.: Fast Analytical Assessment of the Signal Quality in Transparent Optical Networks. Shaker-Verlag, Germany, Sept 2005, ISBN: 978-3832243678
33. Poole, C.D., Tkach, R.W., Chraplyvy, A.R., Fishman, D.A.: Fading in lightwave systems due to polarization-mode dispersion. IEEE Photon Technol. Lett. **3**(1), 68–70 (1991)
34. Kissing, J., Gravemann, T., Voges, E.: Analytical probability density function for the Q factor due to PMD and noise. IEEE Photon Technol. Lett. **15**(4), 611–613 (2003)
35. Gravemann, T., Kissing, J., Voges, E.: Signal degradation by second-order polarization-mode dispersion and noise. IEEE Photon Technol. Lett. **16**(3), 795–797 (2004)
36. Foschini, G.J., Nelson, L.E., Jopson, R.M., Kogelnik, H.: Probability densities of second-order polarization mode dispersion including polarization dependent chromatic fiber dispersion. IEEE Photon Technol. Lett. **12**(3), 293–295 (2000)
37. Agrawal, G.P.: Nonlinear Fiber Optics, 3rd edn. Academic, San Diego (2001)

References

38. Wang, J., Petermann, K.: Small signal analysis for dispersive optical fiber communication systems. J. Lightw. Technol. **10**(1), 96–100 (1992)
39. Cartaxo, A.V.T.: Cross-phase modulation in intensity modulation-direct detection WDM systems with multiple optical amplifiers and dispersion compensators. J. Lightw. Technol. **17**(2), 178–190 (1999)
40. Proakis, J.: Digital Communications. McGraw-Hill, New York (1995)
41. Pachnicke, S., Spälter, S., Reichert, J., Voges, E.: Analytical Assessment of the Q-factor due to Cross-Phase Modulation (XPM) in Multispan WDM Transmission Systems, pp. 61–70. SPIE ITCOM, Orlando (2003)
42. Zeiler, W., Di Pasquale, F., Bayvel, P., Midwinter, J.E.: Modeling of Four-Wave Mixing and Gain Peaking in Amplified WDM Optical Communication Systems and Networks. IEEE J. Lightw. Technol. **14**(9), 1933–1996 (1996)
43. Forghieri, F., Tkach, R.W., Chraplyvy, A.R.: Fiber Nonlinearities and Their Impact on Transmission Systems. Optical Fiber Telecommunications IIIA, Academic, San Diego (1997)
44. Zirngibl, M.: Analytical model of Raman gain effects in massive wavelength division multiplexed transmission systems. IEE Electron. Lett. **34**(8), 789–790 (1998)
45. Huang, J.-H.: Opening Keynote with Jen-Hsun Huang, NVIDIA CEO and Co-Founder. GPU Technology Conference, San Jose (2010)
46. Antona, J.-C., Bigo, S.: Physical design and performance estimation of heterogeneous optical transmission systems. C. R. Physique **9**(9–10), 963–984 (2008)
47. Bononi, A., Bertolini, M., Serena, P., Bellotti, G.: Cross-Phase Modulation Induced by OOK Channels on Higher-Rate DQPSK and Coherent QPSK Channels. IEEE J. Lightw. Technol **27**(18), 3974–3983 (2009)
48. Winter, M.: A Statistical Treatment of Cross-Polarization Modulation in DWDM Systems & its Application. Dissertation, TU Berlin, 2010

Part II
Optical Network Operation

Chapter 5
Dynamic Operation of Fiber Optical Transmission Networks

Abstract This chapter deals with (online) operation of dynamic fiber optical transmission networks. In the first section the network architecture of an optical core network is described, followed by the presentation of the demand model. Afterwards the concept of constraint-based routing considering physical-layer impairments is explained, and a possible implementation is shown. Furthermore, a heuristic for sparse placement of OEO regenerators is outlined. The section is completed by the presentation of simulation results for physical layer impairment (PLI) aware routing and wavelength assignment in a European reference topology. Beyond that the possible reduction of the required number of regenerators by PLI-aware routing is shown. In Sect. 5.4 a concept of considering physical-layer impairments in transmission systems with higher bit rates and novel modulation formats is sketched. Section 5.5 deals with the reduction of the energy consumption in an optical core network by dynamic adaptation to the traffic load changes, which occur on a daily basis. A summary and a discussion are given at the end of this chapter.

It can be observed that in recent years optical networks have been evolving more and more toward flexibility and automatic reconfiguration. At the same time a shift from opaque (where optical–electrical–optical (OEO) conversion is required at each node) to totally or partially transparent networks with optical cross conncets is happening (where OEO conversion is omitted at most nodes). For the above reasons it can be anticipated that providing static and high-capacity pipes will no longer be sufficient to address the demands of emerging dynamic applications [1, 2].

This is why in the future the combination of tunable systems and automatic processes will make networks more flexible to better adapt to changing traffic demands. Network reconfiguration will be possible in the physical—optical transport—layer without many human interventions, ideally fully automatically. A reference architecture for the control plane of such an automatically switched optical network (ASON) has been published by the ITU-T in recommendation G.8080 [3], and the Internet Engineering Task Force (IETF) has developed the Generalized Multi-Protocol Label Switching (GMPLS) paradigm [4], which is a control framework for establishing various types of connections including lightpaths in Internet protocol (IP) based networks. These specifications include signaling protocols to automate control of the optical network and enable features such as discovery of the network topology and the network resources. Beyond these functional features there still exist several technical challenges e.g. the consideration of physical layer impairments to find the optimum route associated to a connection request or demand in (almost) real-time.

The energy consumption of the IT infrastructure is another topic that has shifted into the focus of international research activities recently. The energy consumption is growing as a result of increasing numbers of users and data volume. In access networks (connecting to the end user) the power consumption scales more or less linearly with the number of users due to the deployment of network termination equipment at each subscriber dissipating several Watts per household (relatively independent of the data rate). In the aggregation and core networks the energy consumption increases proportionally with the data volume as network elements have to scale with the growing demands. Today the access sector is dominating the power consumption; however, the energy requirement of the core sector will rise significantly in the future because the data volume is increasing more quickly than the number of users in the saturated telecommunication markets of the industrialized nations. Thus innovative concepts for increasing the energy efficiency in the core network are needed.

5.1 Network Architecture

An optical core network is typically built in a meshed topology composed of nodes and links as shown in Fig. 5.1. The link configuration is similar to the one outlined in the previous chapters. At the beginning of a link in many cases a dispersion pre-compensation fiber (pre-DCF) is deployed. Along the link fiber spans of approximately 80–120 km lengths are cascaded. Periodically—after each span—the signal is re-amplified by an EDFA, which often includes a dispersion compensation module as midstage device.

In the physical layer routing is carried out by optical cross connects (OXCs). Depending on the architecture, an OXC can switch transparently each wavelength of a WDM signal individually and independently from other wavelengths (compare Sect. 2.6). Furthermore, selective wavelengths may be dropped or added at a node. In the case of partially transparent (also termed translucent) optical networks an OEO regenerator bank may be connected to an OXC (Fig. 5.2) to restore the ideal signal quality and to enable longer transmission distances. It shall also be mentioned that transparent switching on a wavelength level may impose additional requirements on the robustness of the network against transients. It is mandatory to

5.1 Network Architecture

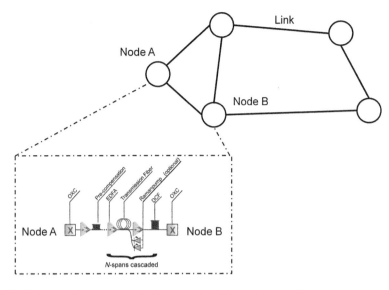

Fig. 5.1 Example of a transparent optical network infrastructure and the components deployed along a link

Fig. 5.2 Node architecture with optional OEO regeneration

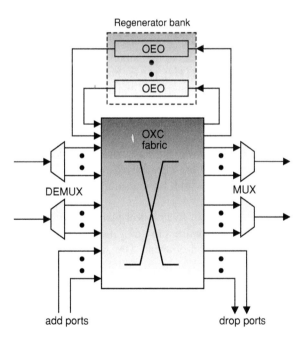

suppress undesired channel power changes efficiently, which may be stemming from e.g. EDFA gain dynamics or changing SRS spectral tilts. For more information on this topic the reader is referred to e.g. [5] and [6].

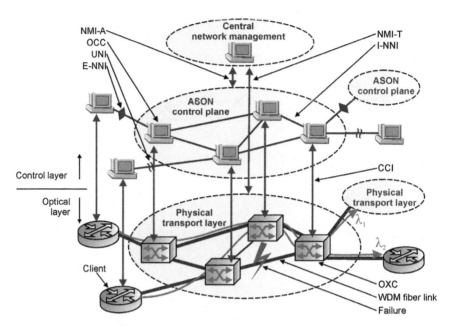

Fig. 5.3 Architecture of an automatically switched optical transport network. It is divided into the physical transport layer, the ASON control plane and the central network management plane

The architecture of an automatically switched optical network (ASON) as proposed by the ITU-T [3] is shown in Fig. 5.3. It is a client–server architecture with well-defined interfaces that allows clients to request services from the optical network. The ASON architecture is divided into three separate planes: the (physical) transport plane, the ASON control plane and the central network management plane.

The lowest layer, the physical transport plane, contains the transport network elements, which are e.g. optical switches (OXCs) and links. End-to-end connections are setup within the transport plane under the control of the ASON control plane.

The purpose of this control plane is to give service providers better control of their networks, while providing fast and efficient circuit setup. It should be reliable, scalable and efficient. The ASON control plane defines a set of interfaces, which are described in the following [3]. Each switch in the optical layer (i.e. OXC) has a counterpart in the control plane, namely an optical connection controller (OCC). Both are connected via the CCI, the connection controller interface. The CCI must support two essential functions: adding and deletion of connections as well as query of the port status of the switches. OCCs are connected in the ASON control plane via the interior network node interface (I-NNI). The two main aspects of I-NNI are signaling and routing. It can be based for example on the IP or GMPLS protocols. For the communication with other networks or domains, the external network node interface (E-NNI) can be used, which offers the same functions as the user network interface (UNI) and some additional features such as exchange of reachability information between different domains (using e.g. the border gateway protocol). The UNI allows ASON clients to create, modify and delete connections. Furthermore, the status may be inquired. Data in the ASON control plane may be transported in a separate network. This has the advantage that the network management can still control the OCCs, if the physical layer is malfunctioning.

The top layer of the ASON architecture is the central network management plane. It is responsible for managing the control plane as well as for fault management, performance management and accounting. This is why connections to both, the ASON control plane and the physical layer, are needed. The network management interface A (NMI-A) provides access to the ASON control plane. In addition, the central network management may directly communicate with the transport plane via the NMI–T. Via the NMI e.g. monitoring information may be passed from the different components to the central network management. Also a feature called "auto-topology-discovery" (ATD) may be implemented. ATD enables to transmit the parameters of new network elements to the network management automatically in this way permitting an automatic adaption to the new configuration. Apart from the provisioning of new connections or the reconfiguration of existing ones, restoration and protection switching are controlled by the network management.

In Sect. 5.3 the aspect of routing and wavelength assignment (RWA) is unfolded in more detail. Efficient RWA is one of the key elements for utilizing the transport infrastructure in an optimum fashion. For transparent networks it is furthermore inevitable to consider the physical-layer performance during the RWA process (so called physical-layer impairment aware constraint-based routing), if the maximum amount of transparency should be used requiring real-time algorithms for the assessment of the signal quality.

5.2 Demand Model

After the network architecture has been described, the traffic offered to the network remains to be defined. While many realistic traffic models for services on higher layers exist, it can currently not be reliably defined when and where connection requests (or demands) in future optical dynamic networks occur [7].

For the arrivals of the connection requests a Poisson process with negative exponentially distributed interarrival time T_A is typically chosen. The probability density function of such an exponential distribution is given by

$$f(x) = \lambda \exp(-\lambda \cdot x), \quad x > 0 \tag{5.2-1}$$

where λ is called the rate parameter, and $1/\lambda$ is the mean.

Fig. 5.4 Schematic of the dynamic connection arrivals and holding times

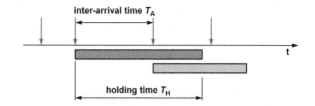

The Poisson process matched telephone traffic well, where a high number of subscribers act independently. Its direct transformation to the traffic in optical networks, however, is not straightforward, especially as future user behavior is yet unknown. The traffic mix in the optical layer comprises different traffic types such as bursty IP traffic, streaming applications but also legacy telephone calls making the traffic model highly complicated. In most publications the Poisson arrival process is best practice at the moment because of its memory-less property. This allows modeling traffic demands where the operator does not know a priori when a connection will be setup.

For the connection holding time T_H also a negative exponential distribution is chosen in most publications. Similar to the argumentation for the interarrival times the exponential distribution is justified for dynamic traffic, where a network operator does not know when a connection will be torn down (Fig. 5.4).

If both interarrival and holding times are exponentially distributed with means $1/\lambda$ and $1/\mu$, respectively, the average traffic volume is defined as λ/μ and measured in the unit "Erlang". Consequently, the traffic load can be increased in a simulation either by decreasing the interarrival time or by increasing the holding time. In the following this is also denoted by a demand scaling factor S describing the increase of the traffic load compared to the initial state with $\lambda = \mu = 1$.

A dynamic traffic matrix is a common way to express the offered traffic between node pairs. As it is quite difficult to generate a traffic matrix with realistic parameters we follow in our studies the procedure suggested in [8, 9]. The basic idea of this approach is to separate traffic into voice, so-called transaction data traffic (data traffic except Internet traffic) and IP traffic to derive corresponding traffic demands from published statistical data. In this model it is assumed that the traffic between major cities i and j depends on the total population of both cities, the number of non-production business employees and the number of Internet hosts in each city, as well as the distance between the two cities of interest. With this information traffic matrices for a set of reference networks can be generated and have already been published in [10]. To take into account the role of internet exchanges (IXP) the above model may be adapted as shown in [11].

The traffic matrix structure described above refers to overall traffic between two node pairs and not to individual demands. It has to be discretized to portions of wavelength demands, if wavelength granularity studies are desired. In this way some 8,000 connection requests are generated for the (large) pan-European reference topology (as shown in Fig. 5.7). These are distributed randomly over the simulation time by the dynamic connection arrival process described above. For a sufficiently large data set—and small confidence intervals—50 to 100 different realizations should be simulated, and the results should be averaged over all simulations.

At the beginning of a simulation, the network load is typically zero. This is why a "warm-up period" should be included in the simulations until a stable operation point is reached [12]. In our simulations the warm-up period expired after 95% of the desired network load has been reached, and results within this period are excluded from the steady-state analysis.

5.3 Constraint-Based Routing and Regenerator Placement

Optical networks of pan-European or trans-American dimensions exhibit path lengths of several thousand kilometers. This makes it essential to consider the signal quality in the routing process because of potentially very long transparent paths, where physical layer degradation effects accumulate and may deteriorate the signal below a required threshold. Such an approach is called constraint-based routing (CBR) or physical-layer impairment (PLI) based routing and has been a subject of intensive research in the last several years (a good summary can be found in [13]).

In the following the approach that we used for the (almost) real-time assessment of the signal quality is outlined. The accuracy of our models has already been discussed in [14]. At this point it shall be pointed out that especially the (complex) models for the nonlinear multichannel fiber degradation effects have been verified experimentally by recirculating loop experiments showing excellent agreement [15–17].

In this section first it is explained how the results for the assessment of the individual physical degradation effects can be aggregated to a single figure of merit (Q-factor) extending the approach outlined in Sect. 4.3. Subsequently, different CBR algorithms are described, and the regenerator placement heuristic is outlined. Finally, results for reducing the blocking probability of dynamic wavelength demands and the required number of OEO regenerators by impairment aware CBR are shown.

5.3.1 Assessment of the Signal Quality by a Single Figure of Merit

In the following we consider the dominant (related to 10 Gb/s NRZ-OOK signals) physical-layer degradation effects only, if not stated otherwise. These dominant effects comprise ASE-noise, PMD, accumulated GVD and filter cross-talk as linear degradation effects, and XPM, FWM and SPM as nonlinear degradation effects.

Other effects can be included in this approach easily as an OSNR—or Q-factor penalty, if desired. An OSNR- or Q-factor penalty means that the respective value (OSNR or Q-factor without this effect) is reduced by a certain amount (called penalty) stemming from the consideration of the additional effect. We assume that a centralized database is available in the network management system (NMS) containing the system parameters needed for the assessment of the degradation effects as discussed in [18]. The individual degradation effects are analyzed separately as already suggested in Sect. 4.3 because in this way a faster—in many cases analytical—estimation of the signal quality is possible.

The procedure of accumulating the different figures of merit (FOM) for the individual degradation effects to a single FOM is as follows. In a first step the Q-factor of a signal impaired by ASE-noise is determined as already shown before in Sects. 3.1.8 and 4.3, respectively,

$$Q_{\text{noise}} = \sqrt{\frac{f_{\text{opt,ref}}}{2 \cdot f_{\text{el}}} \cdot \text{OSNR}} \approx \frac{\mu_1}{\sigma_{1,\text{ASE}}}, \qquad (5.3\text{-}1)$$

where $f_{\text{opt,ref}}$ is the optical filter bandwidth (here: $f_{\text{opt,ref}} = 12.5$ GHz) and f_{el} the (double sided) electrical filter bandwidth (here: $f_{\text{el}} = 14$ GHz, optimized for a 10 Gb/s transmission system). The OSNR at the end of a certain path can be calculated easily, if the amplifier noise figures of the EDFAs along the route are known (e.g. from the database in the NMS).

At this point, PMD induced degradations can be included by the approach shown in [19] as an OSNR penalty. For this purpose the average DGD values for the different links have to be known. Service providers such as Deutsche Telekom AG have measured these parameters for their entire fiber plant. Thus it is assumed that they are available from the NMS for each link in the network.

Afterwards the Q-factor, which has been determined for an entire path—from the origin to the final destination node—is converted to logarithmic units, and Q-factor penalties for the other effects (XPM, FWM, SPM, filter crosstalk) are calculated and subtracted.

$$Q_{\text{Penalty, dB}} = 20 \cdot \log Q_{\text{Penalty, linear}} \qquad (5.3\text{-}2)$$

For the nonlinear multichannel effects of XPM and FWM the Q-factor penalties are given by calculating the ratio of the Q-factor including nonlinear degradation and the Q-factor without the regarded nonlinear effect. Because it is assumed that XPM and FWM do not change the average signal power but only increase the variance of the '1'—level the penalties can be obtained from the following equation [20].

$$Q_{\text{Penalty, linear}} = \frac{Q(\sigma^2_{\text{nonlinear effect}})}{Q(\sigma^2_{\text{nonlinear effect}} = 0)} = \frac{\sigma_{1,\text{ASE}}}{\sqrt{\sigma^2_{1,\text{ASE}} + (R \cdot 2P_{\text{rec}})^2 \sigma^2_{\text{nonlinear effect}}}} \qquad (5.3\text{-}3)$$

5.3 Constraint-Based Routing and Regenerator Placement

In Eq. 5.3-3 the variance $\sigma^2_{\text{nonlinear effect}}$ of the detected mark symbols due to XPM and FWM (compare Eqs. 4.3-18 and 4.3-23) is used. Furthermore, a variance of the detected '1's due to noise $\sigma^2_{1,\text{ASE}}$ is introduced, which can be calculated easily from Eq. 5.3-1, if ASE noise on the '0' symbols is assumed to be small. The second summand in the denominator of Eq. 5.3-3 stems from fiber nonlinearity and accounts for the increase of the signal power variance due to nonlinearities. In this term R stands for the responsivity of the photo diode and P_{rec} for the average received signal power. The advantage of such a procedure is that the noise penalty is modeled for the individual wavelength (or WDM channel) from the source node to the destination node. This enables to track the ASE noise correctly, which has been accumulated along the path in the individual channel. The penalties due to the multichannel degradation effects can be modeled either as fixed precomputed (worst-case) penalties for each link ("offline routing") assuming a fully-loaded system or based on the current network status ("online routing"). The use of pre-computed worst-case penalties is especially attractive as nonlinear degradations caused by FWM or XPM tend to saturate for an increasing number of neighboring WDM channels.

If also the signal degradations due to nonlinear intrachannel impairments (i.e. SPM) shall be included, these are preferably computed numerically by single channel SSFM simulations. The reason is that for the assessment of SPM-induced signal degradations so far no accurate and universally applicable analytical models are available. On the other hand, numerical simulations for single channel effects do not require a long computational time because a rather small number of split-steps is needed (low launch power), and also a relatively small number of samples per bit can be chosen (corresponding to the narrow spectral range and assuming analytical noise modeling as shown in Fig. 3.1, left). Furthermore, SPM-induced penalties may be pre-computed because they do not change with the number of active connections (i.e. current WDM channel count). As SPM degradations cannot be separated from other (linear) intrachannel impairments the simulation results will always include the impairments from (residual) group-velocity dispersion (GVD). The execution time of a numerical simulation for a single channel is much less than a second for a single span on a state-of-the-art desktop computer. The computational time may be optimized further, if the split-step length is increased, however reducing the accuracy of the results. We suggest a maximum admissible nonlinear phase shift of 5 mrad between two split-steps. The proposed SSFM parameters allow estimating a Q-penalty due to intrachannel impairments in the matter of a few seconds for a typical path in the network.

For the assessment of (linear) crosstalk penalties occurring in OXCs the following approach can be used. We assume a node architecture as shown before in Fig. 5.2. In transparent OXCs two different forms of crosstalk may occur: intraband- and out-of-band crosstalk. The former will occur, if a channel is dropped with insufficient isolation, and a certain amount of power remains at the wavelength. Consequently, crosstalk will arise, if a new channel is added at the same frequency. Out-of-band crosstalk is caused by the neighboring channels due to imperfect filter edge steepness. Both forms of crosstalk depend very much on the architecture of the OXCs. They must allow optical pass-through for express channels as well as add/drop for local traffic. Because many different designs for OXCs have been suggested in the literature and are deployed in real networks, in this book a Q-penalty of 1 dB due to crosstalk has been assumed for each OXC included in the transmission path to keep the results generic. The penalties are added, if several OXCs are cascaded along the path leading to a worst-case approximation of the crosstalk. In [21] Q-penalties from filter concatenation with different crosstalk levels have been investigated in great detail confirming our assumption of a 1 dB Q-factor penalty per passed OXC (compare Fig. 23 in [21]). The generic Q-penalty can be exchanged easily by a more accurate calculation of the linear crosstalk, if desired.

5.3.2 Physical Layer Impairment Aware Routing Algorithm

As mentioned above in transparent optical networks routing and wavelength assignment (RWA) are closely linked. Wavelength continuity is an important additional constraint, as in transparent networks wavelength conversion is not possible, and a continuous (free) wavelength has to be found for the entire transmission path (at least if no OEO regenerators are available) of a new connection.

In this book three different routing algorithms will be presented. The simplest one is based on a shortest path algorithm taking the transmission distance as edge weight combined with the constraint of wavelength continuity along the entire desired transparent path. We assume a two step procedure in which first candidate paths are chosen, and afterwards the wavelength continuity is checked for these potential paths. For this purpose the algorithm starts with computing the k-shortest (link disjoint) paths (here: $k = 3$). Afterwards it is checked, whether the maximum transparent distance constraint of the deployed transmission system is met, and only the candidate paths meeting this criterion remain. In the second step it will be checked, if a free continuous wavelength is available, starting with the shortest candidate path. If a free wavelength is available, one wavelength will be selected in a first-fit manner (starting with the lowest frequency). If a regenerator pool is available along the calculated path, wavelength continuity will only be required for the transparent sub-paths. If the first candidate path does not meet the wavelength continuity constraint, the second and then the third path will be tested. If all candidate paths cannot be setup, the connection request will be blocked.

Furthermore, we have investigated CBR algorithms including physical-layer impairments based on either worst-case physical transmission penalties ("offline routing") or on the actual signal degradation occurring in the network with the current load ("online routing"). Obviously, the latter approach is computationally more challenging because the transmission quality has to be calculated "on-the-

5.3 Constraint-Based Routing and Regenerator Placement

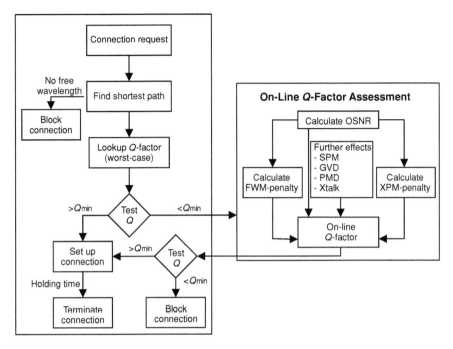

Fig. 5.5 Flowchart of the online CBR algorithm simplified for only one available candidate path

fly" when a connection request occurs. The computational time of the online routing approach is mainly limited by the calculation of the XPM degradation (compare Sect. 4.3) growing with the square of the number of spans and the square of the number of channels. In the offline approach all penalties can be pre-computed. This is why only in the case of a rejection of a connection based on the worst-case penalties "online routing" will be started for this connection request to keep the computational time as low as possible. The more precise "online routing" can determine in this case whether the desired connection really has to be rejected or the current traffic situation will permit to set up the connection request.

A flow chart for the "online routing" algorithm is shown in Fig. 5.5. It starts with finding the k-shortest paths (this time taking the worst-case signal degradation in terms of a Q-factor penalty as edge weight instead of the actual length) between the source and destination nodes. For all possible combinations of start and destination nodes, these paths can be pre-computed and stored in a lookup table. Thereafter the wavelength continuity along the first transparent candidate path is checked. If no free (continuous) wavelength can be found, the next candidate path will be selected. The connection request will be blocked, if no further candidate path is available. In the subsequent step the Q-factor at the destination node is evaluated as outlined in the previous section taking into account the worst-case degradation for a fully loaded system. These worst-case Q-factors can also be pre-computed. Subsequently it is tested, whether the Q-factor of the desired path lies above the required threshold (e.g. $Q = 10$ dB before FEC). If this is the case, the connection will be established. If not, the next candidate path will be checked regarding wavelength continuity and Q-factor threshold requirements. In the "offline routing" scenario the connection request would be rejected, if no more candidate paths were available.

If, however, the online routing algorithm is selected, further assessment of the signal quality will be performed, in this case beginning with the first candidate path and based on the current traffic situation. If this second check yields that the Q-factor is higher than the threshold (e.g. due to a low number of active wavelengths and a resulting low nonlinear degradation), the connection will be setup. Otherwise it is ultimately rejected. We assume that once established connections must not be interrupted (e.g. due to an existing service level agreement). This is why a connection, which is setup with the online approach, is added to a special list of online connections.

If a new connection request occurs, which has any link in common with a connection on this list of online established connections, it has to be checked, whether provisioning the new connection request will affect the online established connection (i.e. degrade the connection by nonlinear crosstalk). If our calculations indicate, that the new connection request would significantly deteriorate any existing connection established by the online routing algorithm (e.g. below the required Q-factor of 10 dB), the new connection request cannot be setup on the presumed route. In that case an alternative route needs to be found, which is possible in most cases. Connections, which have been setup by the offline algorithm, do not need to be considered here because they already assume a fully-loaded system.

Concerning the computational time, the offline routing algorithm is very efficient because it is mainly based on lookup tables (for the shortest-paths and the penalties). The online routing algorithm is much more challenging involving calculations with complex numbers and Fourier transforms. However, the routing of a single demand and the calculation of the online penalties can still be performed below a second on a current desktop computer. Typically, only a fraction of the demands will require online assessment of the signal quality because offline routing will already permit setting up the desired connection.

5.3.3 Regenerator Placement Heuristic

Up to this point fully transparent optical networks have been assumed. However, in reality it is difficult to deploy entirely transparent networks because transmission impairments restrict the maximum reach of a transparent lightpath. This is why (at least sporadic) regeneration is needed. But such OEO regeneration imposes high costs and is inflexible as it is modulation format and bit rate dependent. That is why as few regenerators as possible should be deployed. We furthermore assume that regeneration is only available for a limited number of channels (as shown in

5.3 Constraint-Based Routing and Regenerator Placement

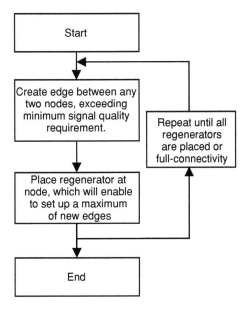

Fig. 5.6 Flowchart of the regenerator placement algorithm

Fig. 5.2) and that the regenerator bank is colorless meaning that any arriving wavelength may be passed to the regenerator. It should be pointed out that the regenerator bank may also be used as a tunable wavelength converter.

In this book the selection of the regeneration sites is based on the estimated physical layer impairments. For a given connection between two nodes, longer than the transparent optical reach, there often exist several possible regenerator locations [22]. However, due to restrictions in capital expenditure (CAPEX) and operational costs (OPEX) the total number of regenerator sites should be kept as low as possible, and it is tried to concentrate regenerators at a limited number of nodes only.

For this purpose a heuristic approach has been employed to place a sparse number of regenerators (inspired by [23]). The procedure is depicted in Fig. 5.6. A new graph is created containing only nodes as defined by the investigated reference network. Then the algorithm tries to set up a complete graph with edges between any combinations of two nodes under the constraint that the minimal Q-factor (e.g. $Q > 10$ dB before FEC) is exceeded. Worst-case Q-factor estimates for all paths in the network can be calculated by the approach outlined in Sect. 5.3.1. If the minimum signal quality is not reached for a specific path, the regarded connection can only be established, if a regenerator is placed at a node along the route. The first regenerator will be placed at that node that will enable to establish the maximum of previously blocked connections, if a regenerator were available there. Afterwards the procedure is repeated and the second regenerator is placed at that location, which will enable to setup the maximum of previously blocked connections and so forth. This is repeated until the maximum allowed number of regenerators has been placed, or full connectivity has been achieved.

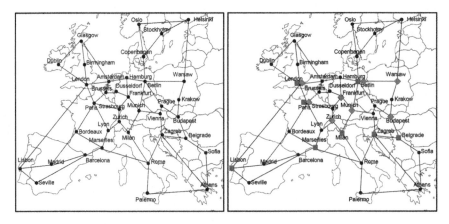

Fig. 5.7 COST266 reference network [10] (*left*) and network with selective regeneration for SSMF (*squares*) and NZDSF (*diamonds*) (*right*)

We did not consider a specific traffic matrix for placing the regenerator pools as in a dynamic network there typically is no (or only limited) knowledge of the traffic matrix a priori, and the demands arise dynamically.

The availability of regeneration has a twofold beneficial effect regarding the blocking probability of dynamic connection requests. First, it enables to setup certain connection requests that were previously blocked due to restrictions in the transparent reach. Second, regenerators have the potential of wavelength conversion. In this way the wavelength continuity constraint is relaxed.

5.3.4 Results

The benefits of CBR including physical-layer impairments will be quantified in the following. As a reference topology the large network of the COST266 project has been used, which is a pan-European network with a total of 37 nodes and 57 links (Fig. 5.7). For this topology demands have been defined based on a population-based model (shown in Sect. 5.2). As a figure of merit, we analyzed the demand blocking probability (BP), which is the ratio of the rejected demands to the total number of offered demands. To facilitate transmission over very long path lengths, pools with regenerators have been placed sparsely at certain nodes (compare Sect. 5.3.3).

Combining intelligent regenerator placement and CBR yields a significantly reduced BP. It is shown that further improvement can be gained by considering the current traffic situation on the links instead of assuming a fully-loaded system. By this means up to five times more traffic can be transported in a network compared

5.3 Constraint-Based Routing and Regenerator Placement

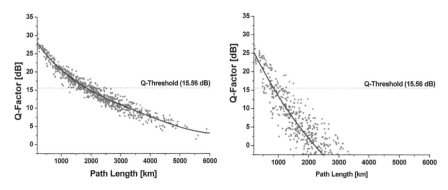

Fig. 5.8 Maximal reach in the investigated network; SSMF (*left*), NZDSF (*right*). Each point represents a path in the COST266 network

to routing with worst-case estimations (for a blocking probability of 3%) [20]. The following exemplary study is based in large parts on our publication [24].

A. Exemplary Network

For the investigations the COST266 reference network (large topology) has been chosen [10]. This network has pan-European dimensions (37 nodes and 57 links) in a mesh topology (Fig. 5.7, left).

Our system simulations have shown that for very long paths in the COST266 network the nonlinear fiber effects of cross-phase modulation (XPM) and four-wave mixing (FWM) cannot be neglected when 10 Gb/s NRZ-OOK modulation is employed (Fig. 5.8). This is why we included the analytical models presented in [14] in our CBR approach to assess XPM and FWM. From Fig. 5.8 an average transparent reach of approximately 1,820 km for SSMF can be observed. However, there are short paths with poor signal quality (minimum 1,280 km) and long paths with good quality (maximum 2,220 km) making an accurate assessment of the actual signal quality desirable. In the COST266 reference network only link lengths and demands have been defined. For the assessment of the signal quality, however, it is essential to know the physical parameters of the links. For this purpose a heuristic approach has been chosen to determine the span and DCF lengths based on realistic assumptions for deployed networks. We assumed a network consisting of SSMF ($D = 17$ ps/nm/km) or NZDSF ($D = 4$ ps/nm/km). Furthermore, the following parameters have been used: launch power $P_{\text{NZDSF}/\text{SSMF}} = 3$ dBm/ch, launch power $P_{\text{DCF}} = -3$ dBm/ch, nonlinearity constant $\gamma_{\text{SSMF}} = 1.37$ $(\text{W} \cdot \text{km})^{-1}$ and nonlinearity constant $\gamma_{\text{NZDSF}} = 2$ $(\text{W} \cdot \text{km})^{-1}$. The span lengths have been generated once from a random process with a Gaussian distribution with a mean value of 80 km and a standard deviation of 5 km. For the DCF modules a granularity of 10 km SSMF-equivalent (i.e. -170 ps/nm) has been assumed yielding an average undercompensation of approximately 85 ps/nm/span along the links. At the nodes a minimal residual dispersion is desirable (under the boundary condition of limited DCF granularity). The investigated modulation format is 10 Gb/s NRZ-OOK. The channel spacing is 50 GHz. A maximum of 80 wavelengths has been assigned to each link. For the EDFAs a noise figure of 5 dB has been assumed. It is important to mention that such a network scenario requires very flexible models because the system parameters may vary from span to span in a deployed network. Furthermore, it is not possible to calculate the transmission performance for a single span and to easily scale it to the total distance because the dispersion management is usually varying.

For the investigations we used a combined regenerator placement and CBR approach as outlined in the last sections. In our scenario chromatic dispersion (CD) does not impose significant penalties due to the modulation format and the dispersion management described above. Furthermore, we assume in this study that an ASON will be built with newer fiber types with low PMD values (i.e. $0.05 \text{ ps}/\sqrt{\text{km}}$). This is why CD and PMD have not been included in these examplary investigations. Each node has been assigned a 1 dB penalty accounting for the signal degradation due to crosstalk [21]. Furthermore, an insertion loss of 10 dB has been associated with each node.

Based on these parameters the regenerator placement has been started. It is tried to minimize the total number of regenerator sites and to concentrate a regenerator pool at a limited number of nodes only (maximizing the number of feasible connections) to reduce costs related to maintenance. In our case a total of eight regenerators have been placed in the network corresponding to approximately 20 % of all nodes. The positions of the regenerators are indicated in Fig. 5.7 (right). Different positions have been obtained for networks consisting of either NZDSF or SSMF. For the latter case full connectivity is achieved with eight regenerators. For NZDSF, however, at least 25 regenerators were needed for full connectivity. To better compare the results to the SSMF simulations we also deployed only eight regenerators in the NZDSF-based network resulting in limited connectivity.

A Poisson process with inter-arrival times of 1 time units for the arrivals of the connection requests and negative exponential distributions with a mean of 1 time units for the holding times are assumed. Thus, in the initial state the offered load (i.e. the product of mean holding time and arrival rate) is 1. Furthermore, the original demands defined in COST266 [10] are multiplied by a linear demand scaling factor S (as also used by [25]) to assess the blocking probability in the future when the total network traffic has increased leading to lower inter-arrival times because all demands need to be fulfilled in the same total simulation time. We assumed the lowest granularity of demands to be 10 Gb/s (equivalent to a full wavelength).

In all studies unprotected services have been assumed. The wavelengths are assigned by "first-fit". In a first approach a shortest-path (SP) algorithm with wavelength continuity constraint has been used. Afterwards we investigated the two different CBR models based on worst-case physical transmission penalties ("offline routing") or the current network status ("online routing"), which have been presented in Sect. 5.3.2. All investigations have been made with the translucent networks shown in Fig. 5.7 (right). Our simulations show that the SP

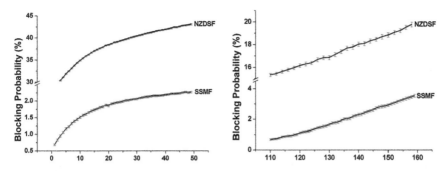

Fig. 5.9 Blocking probability for different routing algorithms; "offline routing" (*left*), online routing (*right*). Depicted are also the 95% confidence intervals

algorithm leads to the highest (more or less constant) blocking probability of approximately 50% in the investigated parameter range ($1 < S < 160$). The reason for this is that certain long paths cannot fulfill the Q-factor requirements and are thus rejected. Both CBR approaches show much lower blocking probabilities (in the case of SSMF-based networks). As expected, the best performance is obtained from "online" CBR. The results are depicted in Fig. 5.9. Because of the lower local dispersion of NZDSF the fiber nonlinearities are more severe resulting in shorter reach and yielding a higher blocking probability for this type of fiber.

5.3.5 Reduction of the Required Number of Electrical Regenerators

Apart from lowering the blocking probability in a dynamic optical network by the use of physical-layer impairment aware routing, it is also possible to reduce the number of required regenerators for a given blocking probability. In the following we will show that the number of OEO regenerators can be reduced by 55% through PLI-aware routing compared to the case with the maximal transparent reach as a criterion (in large parts based on our publication [26]). For a desired blocking probability below 2% regenerators only need to be placed at 12% of the nodes in the investigated trans-American reference topology with a diameter of more than 5,500 km. Furthermore, our studies indicate that the use of PLI-aware routing can also help reducing network resource blocking caused by unavailability of free network elements. The reason is that longer path lengths are feasible in many cases, and detours around highly congested links may be taken.

A. Exemplary Scenario

Figure 5.10 shows the investigated American (US and Canada) reference network [10]. In the used reference network only link lengths and demands have

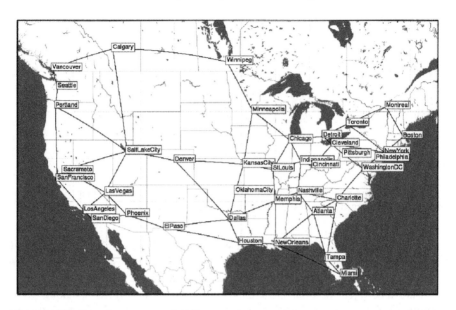

Fig. 5.10 US-CA reference network [10]

been defined. For the assessment of the signal quality, however, it is essential to know the physical parameters of the fiber spans. For this purpose a heuristic approach has been used based on realistic assumptions motivated by deployed networks (mean span length of 80 km and a standard deviation of 5 km). For the DCF modules a granularity of 170 ps/nm has been assumed. The DCF modules have been chosen in such a way that some residual dispersion is left in each span (distributed undercompensation scheme) resulting in less severe accumulation of the nonlinear fiber effects. By this scheme an average undercompensation of 85 ps/nm/span is achieved along the links. At the nodes the residual dispersion is minimized because some channels may be dropped and others may be added.

The distribution of the polarization mode dispersion (PMD) values is based on data from a real network published in [27]. In our simulations the PMD values have been assigned to each fiber span once by a random process based on the distribution shown in published data leading to the numbers shown in Fig. 5.11.

We chose the entire network to consist of SSMF with $D = 17$ ps/nm/km, $\alpha = 0.23$ dB/km, $\gamma = 1.37$ (W·km)$^{-1}$. We assume a channel plan with a maximum of 80 wavelengths per link spaced at 50 GHz (10 Gb/s NRZ-OOK). The EDFAs have been set to $P_{\text{launch}} = 3$ dBm/ch, and a noise figure of $F_o = 5.5$ dB has been assumed.

In a first set of simulations a maximum admissible (transparent) transmission distance of 1,900 km has been assumed (refered to in the following as "reach"). In the next set of simulations all above mentioned physical layer impairments are included by static worst-case penalties ("offline"). Finally, the traffic situation at

5.3 Constraint-Based Routing and Regenerator Placement

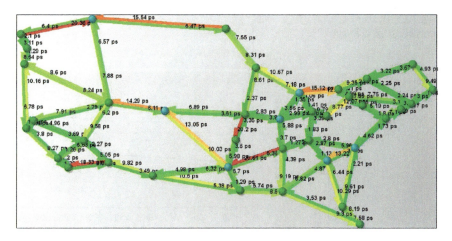

Fig. 5.11 Distribution of the DGD values (*bidirectional links*)

Table 5.1 Regenerator pool locations

Routing scheme	Regenerator pool locations (in descending order)
"Reach" (11 regen.)	Indianapolis, Dallas, Phoenix, Charlotte, San Francisco, Houston, Chicago, New Orleans, Winnipeg, Calgary, Kansas City
"Offline" (7 regen.)	Denver, Dallas, Calgary, St. Louis, Chicago, Atlanta, Charlotte
"Online" (5 regen.)	Denver, Dallas, Calgary, St. Louis, Chicago

the respective point in time is accounted for and the nonlinear degradation effects are assessed according to the current link utilization ("online").

B. Results

The regenerator placement algorithm (compare Sect. 5.3.3) determined the regenerator pool locations as shown in Table 5.1. For a minimum required Q-factor of 10 dB (pre-FEC) and a blocking probability below 2% 11 regenerators need to be placed using the "reach" algorithm, 7 regenerators for the "offline" algorithm and only 5 regenerators when the "online" algorithm is applied. The regenerator placement algorithm determined the same locations for both "offline" and "online" scenarios based on the worst-case physical degradation along the links; however, for the "online" case the two regenerator positions with the lowest priority can be omitted because the "online" CBR algorithm takes a more intelligent routing decision better suited to the physical layer degradation.

In Fig. 5.12 results of our simulations are presented. It can be seen that the (total) blocking probability remains below 2% for a demand scaling factor below 280 in all cases. The demand scaling factor can be used to investigate the performance of the network in the future when the traffic has increased by a factor of

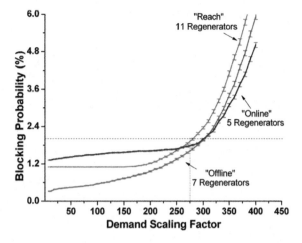

Fig. 5.12 Blocking probability for different scenarios

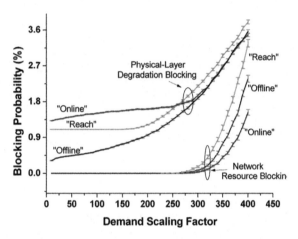

Fig. 5.13 Analysis of the blocking causes

S. Because in the "online" case only 5 regenerators were placed the base blocking probability for a low demand scaling factor is highest. However, it increases less rapidly than the "offline" and "reach" curves, which is favourable when the network load increases.

Furthermore, an analysis of the blocking causes has been carried out. In Fig. 5.13 the curves for physical-layer blocking and network resource blocking are shown.

The sum of both blocking reasons yields the total blocking (as already shown in Fig. 5.12). It can be seen that physical layer blocking is the dominant blocking cause in this example. However, for higher demand scaling factors the lack of free network resources climbs exponentially. It is remarkable that network resource blocking increases more rapidly for "reach" than for "offline" and "online" routing. The reason is that "online" routing allows longer path lengths, and detours around highly congested links are feasible in many cases.

5.4 Extensions to High Bit Rate Systems with Novel Modulation Formats

In this section a potential extension of the PLI-aware routing algorithm to transmission systems with higher channel bit rates and novel modulation formats is presented. The analytical models shown so far (compare Sect. 5.3.1) are suited to 10.7 Gb/s NRZ-OOK transmission systems. As already discussed in Sect. 4.4, however, modifications to next generation transmission systems with novel modulation formats are possible.

PLI-aware routing for 43 Gb/s and 107 Gb/s transmission systems with duobinary or DPQSK modulation will be discussed in the following. Such transmission systems are aiming at cost-sensitive medium (regional) transmission distances and thus cannot rely on sophisticated electrical signal processing at the receiver. For long-distance transport currently coherent polmux QPSK transmission systems with electrical signal processing are state-of-the-art and are already commercially available for 112 Gb/s data rates from some vendors. As a reference topology we study in the following a German network with an average transmission distance of only 485 km. In spite of this relatively low transmission distance the number of passed nodes in this topology is rather high leading to significant additional signal degradation. Furthermore, the sensitivity of the investigated modulation formats towards PMD is demanding. Especially the high variance of the PMD values of different links in the exemplary topology makes the use of PLI-aware routing algorithms desirable.

In the following we show how the previously presented PLI-aware routing models need to be adapted (in large parts based on our publication [28]). We assume that the transmission systems purely use either 43 Gb/s or 107 Gb/s channels, and no mix with legacy 10.7 Gb/s NRZ-OOK channels is present. For such a constellation, SPM will be the dominant nonlinear degradation effect on a standard single mode fiber, and we deliberately neglect impairments stemming from XPM and FWM. Furthermore, PMD strongly limits the maximum transmission distance. Both effects (SPM and PMD) can be estimated with the existing models with only a few modifications.

A. Exemplary Network

We have selected the NOBEL Germany network [10] with 17 nodes and 26 links in a mesh topology with a maximum link length of approximately 400 km (Fig. 5.14, left) as reference network. In this reference topology only the node positions and demands have been defined. This is why we assigned span lengths again by a heuristic approach based on assumptions motivated by deployed networks (mean span length of 80 km and a standard deviation of 5 km) to generate replacements for the missing data. For the DCF modules a granularity of -85 ps/nm has been assumed. The DCF modules have been chosen in such a way that some residual dispersion is left in each span (distributed undercompensation scheme). At the nodes the residual dispersion is minimized. The distribution of the polarization mode

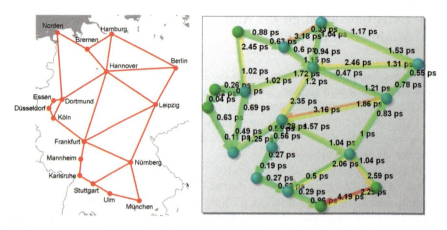

Fig. 5.14 NOBEL Germany 17 node reference network [10] (*left*) and distribution of the mean DGD values for bidirectional links (*right*)

dispersion (PMD) values is based on data from a real German network published in [27]. In that network a considerable amount of legacy fibers with high PMD values above $0.5\ \mathrm{ps}/\sqrt{\mathrm{km}}$ has been deployed. Because these fibers induce a very high OSNR penalty in high bit rate transmission of 43 Gb/s or 107 Gb/s we restricted the distribution of the PMD values to newer fibers with PMD values below $0.5\ \mathrm{ps}/\sqrt{\mathrm{km}}$ (in practice this may be obtained by choosing "better" fibers for transmission on selective spans). This leads to an average PMD value of $0.0825\ \mathrm{ps}/\sqrt{\mathrm{km}}$ in our reference network after the PMD values have been assigned to each fiber span by a random process. The mean DGD values of the bidirectional links are depicted in Fig. 5.14 (right). We chose the entire network to consist of SSMF with $D = 17$ ps/nm/km, $S = 0.056$ ps/nm^2/km, $\alpha = 0.23$ dB/km, $\gamma = 1.37$ (W·km)$^{-1}$. The DCF modules have the following parameters: $D = -102$ ps/nm/km, $S = -0.35$ ps/nm^2/km, $\alpha = 0.5$ dB/km, $\gamma = 5.24$ (W·km)$^{-1}$. We assume a channel plan with a maximum of 40 wavelengths per link spaced at 100 GHz. For the optical filters second-order super Gaussian filters with 3 dB bandwidths of 85 GHz have been used. The electrical filters have 10th order Bessel low-pass characteristics with 40 GHz bandwidth. The EDFAs have been set to $P_{\mathrm{launch}} = 1$ dBm/ch ($P_{\mathrm{launch,\ DCF}} = -1$ dBm/ch), and a noise figure of $F_o = 5.5$ dB has been assumed. As modulation format we used either RZ-DQPSK (50% duty cycle) for 107 Gb/s or 43 Gb/s duobinary, which is generated by electrical filtering with 30% of the bit rate. No (electrical or optical) PMD compensation is used. The investigated scenario exhibits some paths with very high signal quality degradation (and a rather low transparent distance) and other long paths lying well above the required Q-factor of 10 dB (pre-FEC). For 107 Gb/s RZ-DQPSK modulation there e.g. exists a path which only allows bridging 180 km transparently due to high PMD impairments. As will be shown below, however, there exist paths allowing a transparent reach of 642 km showing clearly that routing with fixed transparent reach constraints only will lead to inefficient utilization of transparent reach.

B. Adapted Constraint-Based Routing Algorithm

Our proposed CBR algorithm estimates the transmission quality as follows. For each path in the network the ASE-noise induced OSNR value at the receiver is determined analytically. To this value an OSNR penalty due to PMD is added. The analytical calculation of the (first-order) PMD induced OSNR penalty has been described before in Sect. 4.3.1. This (semi-analytical) PMD calculation requires a pulse form factor, which is set to $A = 1.2$ for 107 Gb/s RZ-DQPSK signals and $A = 2.6$ for 43 Gb/s DB signals. These values are equivalent to a 1 dB penalty for a DGD value of 8 ps in the case of 107 Gb/s RZ-DQPSK and 7 ps for 43 Gb/s DB modulation, respectively. Subsequently, the OSNR value is converted to a Q-factor as shown in Sect. 5.3.1 neglecting nonlinear phase-noise impairments. Afterwards a Q-penalty due to nonlinear intrachannel impairments and group-velocity dispersion (GVD) is added, which can be pre-calculated for each link and each WDM channel. The Q-penalty is determined by a split-step Fourier method approach with a low number of simulated bits (here: 64) and samples per bit (also 64). The (single channel) numerical simulations allow estimating the nonlinear intrachannel and GVD impairments. For this purpose the coupled nonlinear Schrödinger equation is used comprising only the terms for dispersion, attenuation and self-phase modulation (SPM). The execution time of the numerical simulation is less than a second for a single span on a state-of-the-art desktop computer (Intel® Core™ 2 Quad Q6600 with 2.4 GHz clock rate). The computational time may be optimized further, if the split step-size is increased, however reducing the accuracy of the results slightly. We employed a maximum admissible nonlinear phase shift of 5 mrad between two split-steps. The proposed SSFM parameters allow to estimate the intrachannel impairments in the matter of a few seconds for a typical path in the network compared to almost an hour needed for an accurate SSFM based simulation of a WDM system with 40 wavelengths and some 10 spans (which is the length of a typical path).

Our CBR algorithm offers both "offline" or "online" modes. In both cases a k-shortest path algorithm is used to calculate three candidate paths based on precomputed worst-case impairments as edge weights. Afterwards a suitable (continuous) wavelength is searched for, which is assigned from a list of free wavelengths beginning with the longest wavelength. If a regenerator is present along the selected route, it is assumed that wavelength conversion may occur at this point. In offline mode the nonlinear penalties are pre-computed as explained above for each link and wavelength. If a desired path consists of more than one link, the Q-penalties are added for each link (in logarithmic units). This is a worst-case approximation of the transmission quality (similar—but more accurate—to using the nonlinear phase shift for a uniform fiber link as a criterion as proposed by [29].). In a real system the Q-penalty of a concatenation of links is typically smaller because the dispersion map (in our case a distributed undercompensation) has a significant (and beneficial) impact on the accumulation of the penalties. This

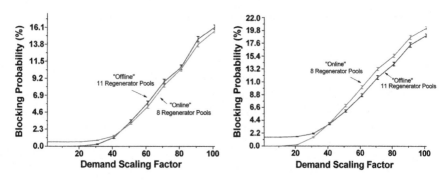

Fig. 5.15 Blocking probability for 107 Gb/s RZ-DQPSK modulation format (*left*) and 43 Gb/s duobinary modulation (*right*)

is why we also implemented an "online" calculation mode. If the signal quality of a candidate path is below a certain threshold (in our case Q = 10 dB, pre-FEC) and the connection request is rejected by the "offline" module, a calculation of the nonlinear Q-penalty for the entire path (comprising several links) is started. This "online" calculation yields a better estimation of the intrachannel and GVD impairments and allows to setup many of the paths, which have been rejected previously. The computational time for an "online" calculation typically lies in the range of only a few seconds (as a rule of thumb approximately 1 s/span in the used configuration). As we did not assume a system with a mix of different modulation formats and channel bit rates (especially with neighboring NRZ channels) it is assumed that XPM impairments are negligible. As shown in the next paragraph the online CBR algorithm allows to reduce the number of required regenerators and also to relax other system requirements such as DCF granularity at the expense of an increased computational effort.

C. Simulation Results

To enable transmission in the NOBEL Germany reference network eight regenerator pools have been placed in the case of 107 Gb/s RZ-DQPSK channels and "online" CBR calculation. To achieve comparable blocking probabilities for the "offline" CBR algorithm three additional regenerator pools have been placed. The regenerator pools enable regeneration of all incoming wavelengths and can also be used as wavelength converters. Furthermore—in offline mode—the residual dispersion at the nodes has been reduced from 0 ± 42 ps/nm ("online") to 0 ± 8.5 ps/nm ("offline"). The simulation results are depicted in Fig. 5.15 (left). The demand scaling factor can be used to investigate the performance of the network in the future when the traffic has grown by a factor of S. Demands are assumed to have full-wavelength granularity. It can be observed that both online and offline routing achieve similar blocking probabilities. However, keep in mind that for offline routing more regenerator pools have been placed and the residual dispersion at the receiver has been lowered. This is why the offline curve starts at a lower blocking probability for small demand scaling factors. Blocking may occur due to signal quality reasons stemming from physical-layer impairments or unavailability of network resources. It is interesting to mention that online routing also decreases network resource blocking because in many cases it allows taking detours to already fully-loaded links. A non-zero blocking probability at a demand scaling factor of $S = 1$ indicates that some paths are not feasible (which could be resolved by adding one additional regenerator pool). However, we assumed a blocking probability of below 2% as acceptable for dynamic network operation and set the minimum number of required regenerators accordingly. In our simulations the average path length is 485 km. The average transparent distance using "online" routing is 242 km versus 175 km in "offline" routing, which is equivalent to increasing the transparent reach by one more span. Even more striking is the difference, if the transparent distance for the longest established path is regarded. This time "online" routing yields 642 km versus 393 km in the "offline" case.

For 43 Gb/s DB transmission similar results have been obtained (shown in Fig. 5.15, right). The results for "online" routing have been computed with the same configuration as for 107 Gb/s RZ-DQPSK channels. For low demand scaling factors the "online" blocking probability is lower than for 107 Gb/s RZ-DQPSK, however the blocking probability is increasing faster than in the 107 Gb/s configuration. The reason may be found in the shrinkage of the dispersion tolerance in the presence of nonlinearity, which is more pronounced for DB than for DQPSK and becomes relevant when alternative routes to already fully loaded links have to be used. In the case of "offline" routing some modifications to the setup have been used. Again three additional regenerator pools have been deployed. Furthermore, the dispersion map has been changed. DB shows a considerable back-to-back penalty due to the V-shape of the eye. This is why we introduced an additional residual dispersion at the receiver of 102 ps/nm for the dropped channels. The average transparent distance using "online" routing is 224 km versus 183 km in "offline" routing. Even more impressive is the difference, if the transparent distance for the longest established path is regarded. This time "online" routing yields 745 km versus 352 km in the "offline" case.

5.5 Improvement of Energy Efficiency

So far the dynamic operation of optical networks has been investigated led by the observation that reconfigurability and flexibility will be some of the drivers of future networks. Recently, the energy consumption of such networks has become a subject of increasing interest. The reasons are twofold. First, energy consumption already makes up a significant portion of the operational expenditure (OPEX). Thus, improving energy efficiency will also reduce the costs of a network operator. Second, the power consumption of the Internet already adds up to 0.75 % of the entire energy consumption of a typical OECD nation [30]. This may appear to be a small fraction, however, as growth rates in the Internet (especially of data traffic) are still very high with approximately 50% per year also the power consumption will rise sharply in the future, if no countermeasures are taken.

Currently, most (>90%) of the power consumption in data communication transport is created in the access area. The customer modem or ONU alone contributes to more than 65% of the total power consumption in access networks, with the DSL modem of each subscriber consuming some 5–10 W [31]. Even for higher access data rates and new technologies (e.g. FTTH) the power consumption of the network termination equipment will stay more or less fixed. In the industrialized world the number of subscribers will not climb significantly anymore, and most people are already connected to the Internet. Thus the energy consumption in network access will only rise slowly in the forthcoming years due to increased backhaul capacity. However, data traffic volume is expected to continue its exponential increase. The energy consumption of the core network is proportional to the transported traffic. This is why the energy consumption of the network core will become dominant in the future, if access data rates are higher than 100 Mb/s [31]. Research efforts should consequently try to improve core network energy efficiency. In this section a study of the energy saving potential of an optical core network is presented especially aimed at reducing the energy consumption during low use phases such as during the night.

Networks have become more and more IP-oriented in the last years. Since the year 2000, data traffic is dominating voice traffic, and also the latter will be transported soon as voice-over-IP (VoIP) in core networks to harmonize the infrastructure. At network nodes IP packets are processed by large (OSI layer 3) routers. For data transport these routers have to communicate with the physical layer (OSI layer 1) equipment. This is referred to as IP over WDM. Historically, for this purpose a large protocol stack was needed (IP over ATM over SONET over WDM) leading to a high protocol overhead. The reason for this architecture to last in many systems up to the present days is that conventional WDM equipment is using SONET (or SDH) to communicate with higher layers [32]. Furthermore, OEO conversion (and electrical routing) is easier to implement than all optical switches. To reduce the protocol overhead (and also associated CAPEX and OPEX), however, several approaches have been suggested in the past [1]. One possibility is to combine the strength of layer 3 routing for small packets, with more efficient switching in layer 2 for high volume traffic and direct end-to-end user flow in layer 1 (bypassing routers) for very large traffic volumes [33]. It is also envisioned to implement direct IP over WDM and to connect the IP routers directly to the physical WDM layer [34]. In the following we assume such a direct architecture for energy consumption analyses. Such an approach is a good starting point as most power is used for IP routers and the underlying physical (or networking) equipment does not contribute much. The goal should be to reduce unnecessary routing in layer 3 and to use optical bypassing of IP nodes as much as possible. Furthermore, it should be investigated how dynamic deactivation of (unused) IP router components (e.g. line cards) can reduce the energy consumption. This is especially important as the traffic volume is varying on a daily basis

5.5 Improvement of Energy Efficiency

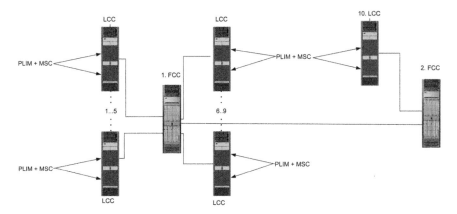

Fig. 5.16 Configuration of a Cisco CRS-1 multishelf system with 10 LCCs and 2 FCCs

with only 25% network utilization in the early morning up to 100% peak traffic load in the evening.

5.5.1 Power Consumption of Deployed Components

Current IP router systems can be configured in a modular way to efficiently suit the required traffic demands of the carriers. Our studies are based on a Cisco CRS-1 multishelf system as an example for an IP core router. A single shelf of such a router offers a switching capacity of 16×40 Gb/s $= 640$ Gb/s, and in full configuration it consumes approximately 10.9 kW. More information can be found directly in the Cisco product specification [35]. A good summary of the power consumption of IP and (optical) physical layer components is also given in [36].

The Cisco CRS-1 system is highly scalable up to (bi-directional) peak traffic of 92 Tb/s (and consuming 1,020 kW). The multishelf system consists of two main components, a line card chassis (LCC) housing the modular services cards (MSCs) and the physical layer interface modules (PLIMs) and a fabric card chassis (FCC) containing the switch fabric cards. The PLIMs can either provide one 40 Gb/s or 4×10 Gb/s optical (short reach) connections (among various other configurations). One LCC can house up to 16 PLIM line cards. Up to 9 LCCs are connected to an FCC in multishelf configuration (compare Fig. 5.16). At one router a total of 8 FCCs may be deployed.

In this book it is assumed that all router components, which are unused, can be shut down immediately. Currently, the deactivation of port groups or line cards is only possible via a local craft terminal or an element management system. For automatic deactivation based on the traffic load an interface to the network management system must be implemented. Various system vendors are working on this functionality. Furthermore, it is desirable that the startup times are reduced, which are currently lying in the region of up to some minutes. In the following it is

Table 5.2 (Peak) Power consumption of selected components [36]

Component	Power consumption [W]
Cisco CRS-1 16 Slot LCC	2,920 (w/o PLIM and MSC) 10,920 (fully configured)
Cisco CRS-1 FCC	9,100
Cisco 40 Gb/s PLIM (STM-256c)	150
Cisco CRS-1 MSC	350
Alcatel-Lucent 40 Gb/s transponder	73
EDFA	8
MRV optical cross connect (OXC)	Max. 700

additionally assumed that at each node at least one LCC is constantly active to enable processing of local traffic. In addition an FCC is needed because it houses stage two of the three-stage Benes switch fabric [37] in the multistage configuration (which can be integrated into the LCC, if a single shelf configuration was deployed).

Apart from the IP router also physical layer elements have a certain power consumption (compare Table 5.2). To this group belong the transponder modules, which are needed to convert the short reach data signals of the PLIMs into long-haul optical signals (and backwards). Furthermore, optical amplifiers (EDFAs) need to be deployed along the transmission path to periodically increase the signal power (as explained in Chap. 2). Last but not least, OXCs may be utilized to enable routing of individual wavelengths in the optical layer.

In our studies we begin with a fixed optical layer meaning that all physical layer components cannot be switched off, even if they become unused. This is a reasonable starting point as associated transients stemming from abrupt reconfiguration of the network are avoided, and also start up times of some active optical components do not need to be considered. In the future also deactivation of physical layer components may be feasible to further reduce the power consumption (as discussed in Sect. 5.5.5).

5.5.2 Grooming

At this point the grooming problem will be explained briefly. This problem has not occurred in the previous sections because always full wavelength demands have been assumed there. As in IP networks, however, we typically deal with traffic granularities well below a wavelength capacity, the aggregation of low bit rate traffic to a high bit rate connection needs to be addressed. In the following studies (Sects. 5.5.4 and 5.5.5) a demand granularity of 0.25 to 42% of a full wavelength is used.

The grooming problem is also referred to as a "bin packing problem" (BPP) in computer science [38]. It is defined as follows: Let there exist n objects with weights $\{a_1, ..., a_n\}$ and k containers with capacity b. An optimum assignment of the objects to the containers is searched for using as few containers as possible and not exceeding the container capacity b. Transferred to our problem the object weights refer to the (sub-granular) data traffic, which needs to be aggregated to wavelength capacities for transport in the optical layer.

It has to be kept in mind that we do not have any a priori knowledge on the upcoming traffic streams so an online algorithm is required. The (online) BPP can be solved for example with a greedy first-fit algorithm, which assigns a low capacity demand to the next available (already provisioned) wavelength with enough free capacity. If not enough capacity is available on existing connections, a new wavelength will be needed. This can be repeated until the maximum installed capacity of a link is used. In the rest of this section the greedy first-fit algorithm is employed for grooming. For a more in depth analysis of online BPP algorithms the reader is referred to e.g. [39].

5.5.3 Approach for Reducing Core Network Energy Consumption

For reducing the power consumption of a core network bypassing of IP routers is a promising approach. As an example, routing in a degree three node with 40 wavelengths on each link will consume more than 100 kW employing a Cisco CRS-1 multishelf system with the above assumptions, whereas routing of optical wavelengths instead only requires one OXC with a few 100 W power consumption.

We start from an opaque network to investigate how optical bypassing can improve the energy consumption. In this configuration at the end of each link an optical-electrical conversion is required, and routing is performed in the IP layer. Afterwards an electrical-optical conversion is used and the data is sent to the neighboring node. Such architecture can still be found in many core networks today. To determine the optimum paths between a pair of nodes a three-shortest paths algorithm is used with a minimum number of hops (passed nodes) criterion. Simulations for different network loads have been conducted for this opaque configuration yielding the (worst-case) power consumption, which is used as a reference for calculating the power savings later.

In a next step partial transparency is introduced in the network by adding optical bypasses, which utilize the same fiber infrastructure (so that the total capacity is kept constant). These bypasses allow individual wavelengths to directly link two IP routers being separated by several intermediate nodes avoiding unnecessary OEO conversions, if desired. The optimum positions can be selected based on the utilization of link tuples (chains of edges) with a certain length. As a starting point for a reasonable bypass length the average path length occurring in a given topology may be used (which is e.g. approximately three for the COST266 European core topology). Subsequently, the utilization of all link tuples with the desired length needs to be analyzed. Utilization is defined in this context as the product of the number of connections and the usage time. It can be calculated from simulations with a random demand model for a specific network load and should be averaged over a sufficiently high number of independent realizations. In real networks utilization of certain paths may be available from the NMS and should be averaged over certain time periods of a day with more or less constant network load.

Using this information the tuple with the highest utilization is chosen as an optical bypass first. In the physical layer this requires deploying OXCs at the intermediate nodes. Consecutively, the optical bypass (represented by a single edge) is added to the logical network topology, and the routing table is updated accordingly. Further optical bypasses may be included by repeating the above steps. In this way more and more optical transparency is added to the network.

A drawback of this method is that possibly network utilization is reduced as grooming will not be possible at intermediate nodes. This effect is especially detrimental for high network loads. In this case the blocking probability is increased [40], which is undesired. Thus optical bypassing is more attractive for phases of medium and low network loads. Furthermore, it has to be kept in mind that reconfiguration of already existing connections may be unfeasible due to service level agreements requiring uninterrupted connections.

5.5.4 Examplary Study

In the following a study of the network energy consumption of a photonic IP over WDM core network is presented (based on our publication [41]). It is shown that utilization of only ten optical bypasses (or virtual links between IP routers) may reduce the dynamic portion (above standby) of the energy consumption by more than 19%. In combination with deactivation of unused network resources this approach enables power savings of 57% in low load periods (e.g. during the night).

A. Investigated Network and Demand Model

For our investigations we have selected the COST266 European core reference topology shown in Fig. 5.17 (left). The network consists of 28 nodes and 41 bidirectional edges in a mesh topology. As before we assumed the span lengths to be Gaussian distributed with a mean value of 80 km and a standard deviation of 5 km. At the beginning of each span an EDFA is placed. Furthermore, it is assumed that OXCs are available at nodes to allow optical bypass. As most systems designed for 80 channels are equipped only partially with transponders we also assume only up to 40 wavelengths per link with a bit rate of 40 Gbit/s per wavelength. In our studies—as a first step—we assume that all components in the optical layer remain active all the time (fixed optical layer) and that all (IP router) line cards and chassis can be switched off, if they do not carry active traffic (flexible IP layer). In this study each node is equipped with a Cisco CRS-1 IP

5.5 Improvement of Energy Efficiency

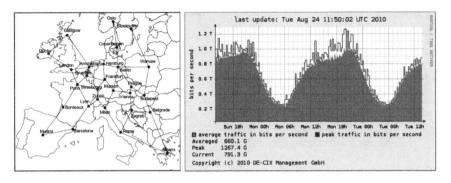

Fig. 5.17 COST266 reference network (*left*) and daily usage statistics of DE-CIX node (*right*) [42]

router—as an example of a large core router—configured in such a way that the peak load of all incoming links can be processed.

Core networks are designed for peak traffic demands [43], however, peak traffic load does not occur permanently. In Fig. 5.17 (right) the traffic load changes during the course of the day of the DE-CIX internet exchange in Frankfurt/Main, Germany, are displayed. The curve on top depicts the peak traffic load (in bit/s), whereas the underlying area depicts the average traffic load (averaged over 15 min intervals). It can be observed that the traffic load shows variations during the course of the day ranging from approximately 25% in the morning to peak values in the evening [44]. We assume that the usage of the DE-CIX node is a good indicator for the utilization of the entire network.

The (static) traffic matrix for the COST266 reference network used in the previous sections is based on a population model. From these data we created a dynamic traffic matrix with 8,564 demands with granularity ranging from 0.25 to 42% of a full wavelength (more details can be found in [45]). In our studies we investigated four different network loads ranging from 25% minimum load to 100% peak load. We simulated a total of 50 random demand realizations for each investigated traffic load. The peak load (100% load) of the investigated network is 560 Erlang. For that value a maximum blocking probability (ratio of rejected to offered demands) of below 2% has been measured. In a real network it can be expected that this blocking probability will not lead to traffic loss as the transport layer protocol would detect network congestion on certain links and would therefore temporarily decrease the data transmission rate [46]. In our study grooming of demands is possible at all IP nodes.

B. Heuristic Approach to Reduce Energy Consumption and Results

In the initial state of our simulation the network is assumed to be totally opaque meaning that the optical connections are terminated after each link. All demands are processed at each node by an IP router. For transporting the demands with a minimum number of hops we used a 3-shortest (link-disjoint) paths algorithm.

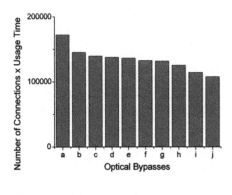

Fig. 5.18 Determination of optical bypasses by highest usage determined for a network load of 140 Erlang

As the IP router is the network element with the highest contribution to the total energy consumption of the network (compare Table 5.2) optical bypassing of IP routers is a promising option to decrease the total power consumption. To investigate the impact of optical bypasses we successively add transparent connections to the initially opaque network. For these transparent connections we chose a length of three links. The reason is that for the investigated topology the average number of hops of a connection is approximately three. Furthermore, for this choice the maximum transparent reach of a 40 Gb/s optical signal—which is roughly between 1,000 and 1,500 km, if several intermediate OXCs are passed—is not exceeded. To select the optical bypasses to be configured we used the following heuristic approach. We analyzed 50 random demand realizations for a given traffic load and generated a list of all occurring continuous link tuples of length three. In a real network the analysis of the network traffic load on the links over a period of several days may be used instead. It is assumed that traffic patterns are very similar and recur on a daily basis. The tuples are sorted by utilization (defined as the product of number of connections times the usage time). The tuple utilized most is added first as an optical bypass to the graph, then the second and so on.

The exact position and the ranking of the different optical bypasses are shown in Fig. 5.18. We assume that existing connections carrying active traffic cannot be switched to an optical bypass (e.g. due to a service level agreement requiring uninterruptable connections). In the following we always used the bypasses determined for a traffic load of 140 Erlang and in the order listed in Fig. 5.18. We only consider ten optical bypasses in this study to keep the number of reconfigurations of the network relatively low.

In the optical layer transparent bypasses can be created by deploying optical cross-connects (OXCs) at the intermediate nodes, and bypasses may be loaded up to the total number of (continuous) wavelengths available (here: 40). We assume that the optical bypasses use the same infrastructure as the original opaque network

5.5 Improvement of Energy Efficiency

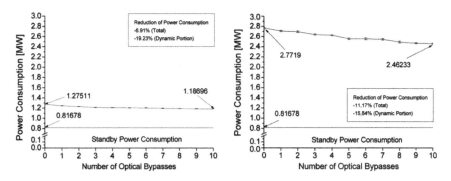

Fig. 5.19 Total power consumption of the investigated topology for 25% network load (140 Erlang) (*left*) and maximum (*right*) traffic load (560 Erlang)

Table 5.3 Reduction of the total power consumption for different traffic loads (standby power consumption: 0.81678 MW)

Network load [Erlang]	Power consumption with 0 bypasses [MW]	Power consumption with 10 bypasses [MW]	Reduction of power consumption (total) [%]	Reduction of power consumption (dynamic portion) [%]
140	1.27511	**1.18696**	6.91	19.23
280	1.7457	1.57327	9.88	18.56
420	2.2793	2.00099	12.21	19.03
560	**2.7719**	2.46233	11.17	15.84

so no additional capacity is added to the network. We furthermore assumed that it is not possible to shut down an IP node entirely because there is always local traffic at each node. With these assumptions a minimum power consumption of the entire network comprising both IP routers and optical layer equipment can be calculated. For the investigated network the standby power consumption is 0.82 MW (Fig. 5.19, Table 5.3). For the peak network load a maximum power consumption of 2.77 MW has been determined. It is obvious that the energy consumption is highest for the opaque case with no optical bypasses. By adding 10 optical bypasses the total energy consumption is decreased by approximately 11% in that case (Fig. 5.19 right, Table 5.3).

If only the variable power consumption (power consumption above standby, which can actually be influenced by optical bypassing of IP routers) is regarded, this part is decreased by 15.8%. For lower network loads the maximum power consumption (which occurs for the totally opaque network scenario with no bypasses) is reduced due to the fact that only line cards and chassis are powered up when needed. In the case of 25% network load (equivalent to 140 Erlang in our studies) a maximum power consumption of 1.28 MW has been calculated. By adding 10 optical bypasses this number can be reduced by another 6.9% (or 19.2%, if only the dynamic portion of the energy consumption is referred to).

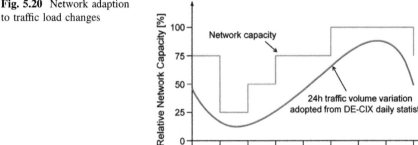

Fig. 5.20 Network adaption to traffic load changes

This leads to a total reduction of the power consumption of 57% compared to the peak network load.

From Fig. 5.19 it can also be observed that the reduction of the dynamic portion of the power consumption is approximately 19% for network loads of 140–420 Erlang when adding 10 optical bypasses. In the case of peak network load (560 Erlang), however, this number is reduced to 15.8%.

This can be attributed to the fact that some links (especially in the center of the network) may be fully loaded and detours to these routes have to be taken leading to a lower utilization of the optical bypasses. This in turn leads to an increased blocking probability, which may be undesired. Also the use of optical bypasses may be limited due to the unavailability of free network resources. The use of load-adaptive optical bypasses is thus preferential in low load scenarios when unused capacity in the optical layer exists.

5.5.5 Reduction of Energy Consumption by Load-Adaptive Operation

As it has already been pointed out the traffic load shows variations in the course of the day ranging from approximately 25% in the early morning to peak values in the evening (compare Fig. 5.17, right). In this section we analyze network energy consumption in a dynamic traffic scenario assuming deactivation of network element subsystems during low load periods and only a limited amount of reconfiguration of the optical layer (e.g. to allow optical bypassing of IP routers) based on the results shown in Sect. 5.5.4. In this way the core network can follow the traffic variations in the course of the day roughly. A low number of reconfigurations (here investigated: every three hours) is desirable to ensure stable network operation and to reduce undesired side effects such as transients due to large channel load changes in the optical layer. We use four different operation points for 25, 50, 75 and 100% traffic load (compare Fig. 5.20).

5.5 Improvement of Energy Efficiency

We consider the same boundary conditions as in the previous section and start our analysis at 100% traffic load. For the case of partially loaded networks we investigate three different options for reducing the power consumption, which are explained in the following.

First we study the case where line cards (and chassis) in the IP router can be deactivated, if they become unused (dynamic deactivation in IP layer). The router solution is assumed to have a module-based sleep mode, where idle modules do not consume power (further details can be found in [47]). In our investigations we furthermore minimize the number of active chassis (i.e. the number of active line card per chassis is maximized). At this point deactivation and activation times should be taken into consideration, which currently can lie in the range of some minutes. As we only assume network reconfiguration every 3 h, however, these startup times do not pose any issues. We did not make any attempts to optimize utilization of specific links or to empty lightly used links entirely because this would require extensive reconfiguration of the network. We additionally assume that it is not possible to shut down an IP node entirely because there is always local traffic at each node. To avoid any issues from dynamic reconfiguration of the optical layer (e.g. power transients), the optical layer remains fixed in this investigation, and all optical transponders remain active during the entire simulation.

To further reduce the energy consumption optical bypassing of IP routers may be introduced (dynamic IP layer & optical bypasses). We assume that optical bypasses use the same infrastructure as the originally opaque network so no additional capacity is added. Dynamic setup of transparent bypasses is facilitated by optical cross-connects (OXCs) at the intermediate nodes. We assume that OXCs are colorless, directionless and contentionless. We investigate the configuration of optical bypasses (virtual links with only 1 hop) of IP routers between up to 10 node pairs depending on the network load. The positions of the bypasses are determined based on the utilization of paths in the network (as described in Sect. 5.5.4). In our simulations the utilization has been averaged over several statistically independent simulation runs (for a given network load), and different to the previous section possible bypass lengths of 2, 3 and 4 have been considered. Utilization of bypasses of length 3 and 4 is weighted by a factor of 2 and 3, respectively, to take into account the higher potential energy savings of longer bypasses. In this second scenario we again assume that all components in the optical layer remain active all the time.

In our third scenario we additionally consider a dynamic optical layer being able to deactivate unused transponders (dynamic IP & optical layers & optical bypasses). This scenario promises the highest energy savings in low load scenarios, however, comes at the expense of a more complex network management and potential transient effects associated to channel load changes in the optical layer. We assume the investigated system to be immune to transients in this third scenario (e.g. due to fast control of EDFA pump powers, etc.).

The power consumption shown in Fig. 5.21 (left) has been averaged over the entire simulation period and all random demand utilizations. It can be seen that the

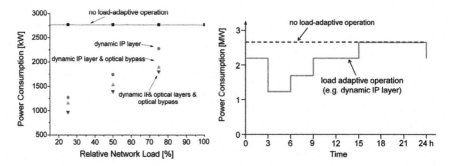

Fig. 5.21 Total network power consumption for different amounts of network load (*left*) and power consumption over time (*right*)

three different options for reducing the network energy consumption all give significant savings, if network load is less than 100%, compared to the case of static network operation. Dynamic operation of the IP layer already leads to savings of 53% for the case of 25% network load. Additional savings (in total 57%) can be gained, if optical bypassing to IP routers is used. The highest energy savings of up to 63% are achieved, if dynamic IP and optical layers are considered and additionally optical bypasses are utilized.

In Fig. 5.21 (right) the power consumption has been plotted over time assuming the traffic load variation already depicted in Fig. 5.20. To increase clarity only the curves for dynamic IP layer operation and for no load-adaptive operation are depicted. If the total daily energy consumption is calculated (the integral below the power consumption curve), savings of up to 17% could be achieved using the dynamic IP layer approach. For additional use of optical bypasses and the most sophisticated scheme with dynamic IP and optical layers as well as the usage of bypasses savings of 23 and 26%, respectively, have been calculated.

5.6 Summary and Discussion

It is envisioned that in the future lightpaths may be requested by the user on-demand [13]. Photonic paths are not currently common service offerings, but they will offer significant infrastructure benefits such as reduced cost, space and power dissipation. To setup such dynamic wavelength demands and to utilize the maximum amount of transparency new physical-layer impairment aware routing algorithms are needed. In this chapter the dynamic operation of fiber optical networks has been investigated.

The chapter started with a presentation of the network architecture of an automatically switched optical network and of the demand model, which has been used in the analyses. Afterwards physical-layer impairment aware routing, which requires estimating the signal quality in real time, and the wavelength assignment

5.6 Summary and Discussion

problem in a transparent network have been addressed. The models used for the Q-factor assessment have already been explained in Chap. 4. In this chapter consequently only some additions have been presented for degradation effects not covered so far, and the aggregation of the different estimates for the individual effects to a single figure of merit has been shown. The worst-case signal degradation along a link has been used afterwards as edge weight in the routing algorithm. To further improve the accuracy of the signal quality assessment, the CBR algorithm also allows considering the current network load and associated (nonlinear) crosstalk between different channels.

As core networks with a reach of several thousand kilometers can exceed the transparent reach of optical transmission systems OEO regeneration is needed at certain locations. Electrical regeneration, however, is costly and should be avoided, if possible. This is why a regenerator placement algorithm has been presented allowing sparse placement of regenerator pools in the network based on estimates of the worst-case signal degradation.

Results have been presented in Sects. 5.3.4 and 5.3.5. It has been shown that considering the current network load enables to significantly lower the blocking probability of dynamic wavelength requests, if links are not fully loaded compared to a CBR algorithm based on worst-case estimates of the signal quality. The inclusion of the estimated signal quality may also be used to decrease the required number of electrical regenerators in a network by 55% (compared to routing with a maximum transparent distance criterion). This has been discussed subsequently based on simulation results for a topology of a trans-American (US-Canadian) network.

As (analytical) modeling of the signal quality needs to be adapted for specific bit rates and modulation formats, possible extensions to higher bit rates and novel modulation formats have been presented in the following section. Simulation results for exemplary transmission systems with channel bit rates of 43 and 107 Gb/s have shown a 38% increase in average transparent distance by the use of the CBR algorithm compared to routing with worst-case signal quality assumptions.

The improvement of the energy efficiency, a relatively new topic in optical networking, has been addressed in the last section of this chapter. The section began with the presentation of the investigated network model and a summary of the energy consumption of the deployed components. Afterwards an approach for reducing the energy consumption has been shown by introducing optical bypasses, which transport more traffic transparently in the optical layer, enabling partial deactivation of unused network resources in the IP layer. As a reference scenario to study the energy saving potential we have chosen a backbone network showing strong traffic load changes on a daily basis with a peak load in the evening and a relatively low utilization of approximately 25% of that value during the night. For optimum energy efficiency it is desirable that the power consumption of the network components can follow this characteristic. Our results have shown that by the investigated approach the energy consumption can be reduced by 57% during low load periods.

It will be interesting to see how dynamic provisioning of wavelength demands will be implemented in future optical transport networks. Until now bandwidth on demand services are very rare (e.g. AT&T's Optical Mesh Service [48]), and generally such services are not yet available for rates of more than 2.5 Gb/s. In a recent post deadline paper of the Optical Fiber Communication Conference [49] an experimental demonstration of impairment-aware control plane schemes has been presented extending the GMPLS control protocols. Up to now it is unclear whether the real-time signal quality estimation will be based on analytical models as shown in this book or a simpler yet more inaccurate criterion such as the nonlinear phase shift may be used [50]. It is also possible to base the routing decision on signal quality information gathered from optical monitoring equipment distributed in the network. The last approach, however, requires significant capital expenditure.

In the USA the Defense Advanced Research Projects Agency (DARPA) has recently started the CORONET program ("Dynamic Multi-Terabit Core Optical Networks: Architecture, Protocols, Control and Management") investigating rapid provisioning in 100 ms or less of future dynamic wavelength services ranging from speeds of 40–800 Gb/s [51]. The project seeks to establish a commercially-viable network architecture. It will be interesting to follow whether these ambitious goals will be fulfilled and which technologies will be proposed.

Also the research on energy efficiency improvements in core networks has just started. Most of the publications so far study possible concepts (compare e. g. [43, 52]) or gedanken network models in order to determine the possible energy savings and minimum energy requirements [53, 54]. Recently the GreenTouch^{TM} consortium has formed with partners from the information and communication technology industry, academia and non-governmental participants. The goal is to transform communications and data networks to reduce significantly the carbon footprint of devices, platforms and networks [55]. By 2015 the network energy efficiency is intended to be increased by a factor of 1,000 from the current level. The future will tell whether these ambitious goals can be achieved, and how long it will really take to replace the current infrastructure.

References

1. Berthold, J., Saleh, A.A.M., Blair, L., Simmons, J.M.: Optical networking: past present, and future. IEEE/OSA J. Lightw. Technol. **26**(9), 1104–1118 (2008)
2. Gladisch, A., Braun, R.-P., Breuer, D., Erhardt, A., Foisel, H.-M., Jäger, M., Leppla, R., Schneiders, M., Vorbeck, S., Weiershausen, W., Westphal, F.-J.: Evolution of terrestrial optical system and core network architecture. Proc. IEEE **94**(5), 869–891 (2006)
3. ITU-T Recommendation G.8080, Nov 2006
4. Mannie, E.: Generalized multi-protocol label switching (GMPLS) architecture. IETF Network Working Group, RFC 3945, Oct 2004
5. Pachnicke, S., Gottwald, E., Krummrich, P., Voges, E.: Transient gain dynamics in Long-Haul transmission systems with electronic EDFA gain control. OSA J. Opt. Netw. **6**(9), 1129–1137 (2007)

6. Pachnicke, S., Gottwald, E., Krummrich, P., Voges, E.: Combined impact of Raman and EDFA transients on Long Haul transmission system performance. European Conference on Optical Communication (ECOC 2007), P074, Berlin, Germany, Sept 2007
7. Schupke, D., Duhovnikov, S., Meusburger, C.: Guidelines for connection-level performance simulation of optical networks. Technical Report, LKN-TR-5, Institute of Communication Networks, TU München, Jan 2010
8. Hülsermann, R., Bodamer, S., Barry, M., Betker, A., Gauger, C., Jäger, M., Köhn, M., Späth, J.: A set of typical transport network scenarios for network modeling. ITG-Conference "Photonic Networks", Leipzig, Germany, May 2004
9. Dwivedi, A., Wagner, R.E.: Traffic model for USA long-distance optical network. Optical Fiber Communications Conference (OFC), TuK1-1, Mar 2000
10. COST 266: Pan-European and Trans-American reference networks, http://sndlib.zib.de
11. Autenrieth, A., Elbers, J.-P., Schmidtke, H.-J., Macchi, M., Rosenzweig, G.: Benefits of integrated packet/circuit/wavelength switches in next-generation optical core networks. Optical Fiber Communications Conference (OFC), NMC4, Los Angeles, Mar 2011
12. Pawlikowski, K., Joshua Jeong, H.-D., Ruth Lee, J.-S.: On credibility of simulation studies of telecommunication networks. IEEE Comm. Mag. **40**(1), 132–139 (2002)
13. Azodolmolky, S., Klinkowski, M., Marín Tordera, E., Careglio, D., Solé-Pareta, J., Tomkos, I.: A survey on physical layer impairments aware routing and wavelength assignment algorithms in optical networks. Elsevier J. Comput. Netw. **53**(7), 926–944 (2009)
14. Pachnicke, S., Reichert, J., Spälter, S., Voges, E.: Fast analytical assessment of the signal quality in transparent optical networks. IEEE J. Lightw. Technol. **24**(2), 815–824 (2006)
15. Pachnicke, S., Hecker-Denschlag, N., Spälter, S., Reichert, J., Voges, E.: Experimental verification of fast analytical models for XPM-impaired mixed-fiber transparent optical networks. IEEE Photonics Technol. Lett. **16**(5), 1400–1402 (2004)
16. Pachnicke, S., De Man, E., Spälter, S., Voges, E.: Impact of the inline dispersion compensation map on four wave mixing (FWM)—impaired optical networks. IEEE Photonics Technol. Lett. **17**(1), 235–237 (2005)
17. Pachnicke, S., Gravemann, T., Windmann, M., Voges, E.: Physically constrained routing in 10 Gb/s DWDM networks including fiber nonlinearities and polarization effects. IEEE J. Lightw. Technol. **24**(9), 3418–3426 (2006)
18. Martinez, R., Pinart, C., Cugini, F., Andriolli, N., Valcarenghi, L., Castoldi, P., Wosinska, L., Comellas, J., Junyent, G.: Challenges and requirements for introducing impairment-awareness into the management and control planes of ASON/GMPLS WDM networks. IEEE Comm. Mag. **44**(12), 76–85 (2006)
19. Kissing, J., Gravemann, T., Voges, E.: Analytical probability density function for the Q factor due to PMD and noise. IEEE Photonics Technol. Lett. **15**(4), 611–613 (2003)
20. Pachnicke, S., Paschenda, T., Krummrich, P.: Assessment of a constraint-based routing algorithm for translucent 10 Gb/s DWDM networks considering fiber nonlinearities. OSA J. Opt. Netw. **7**(4), 365–377 (2008)
21. Vasilyev, M., Tomkos, I., Mehendale, M., Rhee, J.-K., Kobyakov, A., Ajgaonkar, M., Tsuda, S., Sharma, M.: Transparent ultra-Long-Haul WDM networks with broadcast-and-select OADM/OXC architecture. IEEE J. Lightw. Technol. **21**(11), 2661–2672 (2003)
22. Simmons, J.M.: Network design in realistic all-optical backbone networks. IEEE Commun. Mag. **44**(11), 88–94 (2006)
23. Chen, S., Raghavan, S.: The regenerator location problem. Working Paper, Smith School of Business, University of Maryland (2006)
24. Pachnicke, S., Paschenda, T., Krummrich, P.: Physical impairment-based regenerator placement and routing in translucent optical networks. Optical Fiber Communications Conference (OFC 2008), OWA2, San Diego, USA, Feb 2008
25. Schupke, D.A., Jäger, M., Hülsermann, R.: Comparison of resilience mechanisms for dynamic services in intelligent optical networks. Design of Reliable Communication Networks (DRCN), Oct 2003

26. Pachnicke, S., Krummrich, P.: Reduction of the required number of electrical regenerators by physical layer impairment aware regenerator placement and routing. European Conference on Optical Communication (ECOC 2008), Brussels, Belgium, Sept 2008
27. Breuer, D., Tessmann, H.-J., Gladisch, A., Foisel, H. M., Neumann, G., Reiner, H., Cremer, H.: Measurements of PMD in the installed fiber plant of Deutsche Telekom. IEEE LEOS Summer Topical Meetings, MB2.1, July 2003
28. Pachnicke, S., Luck, N., Krummrich, P.: Novel physical-layer impairment-aware routing algorithm for translucent optical networks with 43 Gb/s and 107 Gb/s Channels. IEEE International Conference on Transparent Optical Networks (ICTON 2009), Ponta Delgada, Portugal, June 2009
29. Antona, J. C., Bigo, S., Faure, J.-P.: Nonlinear cumulated phase as a criterion to assess performance of terrestrial WDM systems. Optical Fiber Communications Conference (OFC), WX-5, Anaheim, Mar 2002
30. Hinton, K., Baliga, J., Ayre, R., Tucker, R. S.: The future Internet—an energy consumption perspective. IEEE OptoElectronics and Communications Conference, Hong Kong, 2009
31. Baliga, J., Ayre, R., Hinton, K., Sorin, W.V., Tucker, R.S.: Energy consumption in optical IP networks. IEEE J. Lightw. Technol. **27**(13), 2391–2403 (2009)
32. Ramaswami, R., Sivarajan, K.N., Sasaki, G.H.: Optical Networks–A Practical Perspective, 3rd edn. Morgan Kaufmann, Burlington (2010)
33. Magill, P.: Core photonic networks—where are things heading. European Conference on Optical Communications (ECOC), 4.6.1, Vienna, Sept 2009
34. Mukherjee, B.: Optical WDM Networks. Springer, Berlin (2006)
35. Cisco CRS-1 product specification. www.cisco.com/en/US/prod/collateral/routers/ps5763/prod_brochure0900aecd800f8118.pdf. Version 2010
36. Idzikowski, F.: Power consumption of network elements in IP over WDM networks. TKN Technical Report Series TKN-09-006, Telecommunication Networks Group, Technical University Berlin, 2009
37. Benes, V.E.: On rearrangeable three-stage connecting networks. Bell Syst. Tech. J. **41**, 1481–1492 (1962)
38. Vazirani, V.V.: Approximation Algorithms. Springer, Berlin (2001)
39. Borodin, A., El-Yaniv, R.: Online Computation and Competitive Analysis. Cambridge University Press, Cambridge (1998)
40. Monti, P., Wiatr, P., Jirattigalachote, A., Wosinska, L.: Trading power savings for blocking probability in dynamically provisioned WDM networks. IEEE International Conference on Transparent Optical Networks (ICTON 2010), Munich, June 2010
41. Pachnicke, S., Kagba, H., Krummrich, P.: Load adaptive optical-bypassing for reducing core network energy consumption. ITG-Conference "Photonic Networks", Leipzig, Germany, May 2011
42. DE-CIX German Internet Exchange (www.decix.de)
43. Lange, C., Kosiankowski, D., Gerlach, C., Westphal, F.-J., Gladisch, A.: Energy consumption of telecommunication networks. European Conference on Optical Communications (ECOC), 5.5.3, Vienna, Sept 2009
44. Lange, C., Gladisch, A.: Energy efficiency limits of load adaptive networks. Optical Fiber Communications Conference (OFC), OWY2, San Diego, Mar 2010
45. Pachnicke, S., Krummrich, P.: Constraint-based routing in path-protected translucent networks considering fiber nonlinearities and polarization mode dispersion. SPIE Asia-Pacific Optical Communications Conference (APOC 2008), Hangzhou, China, Oct 2008
46. Hülsermann, R., Lange, C., Kosiankowski, D., Gladisch, A.: Analysis of the energy efficiency in IP over WDM networks with load-adaptive operation. ITG-Conference "Photonic Networks", Leipzig, Germany, May 2011
47. Hasan, M.M., Farahmand, F., Jue, J.P.: Energy-awareness in dynamic traffic grooming. Optical Fiber Communications Conference (OFC 2010), paper OWY6, San Diego, Mar 2010

48. Doverspike, R.: Practical aspects of bandwidth-on-demand in optical networks. Optical Fiber Communications Conference (OFC 2007), Panel on Emerging Networks, Service Provider Summit, Anaheim, CA, Mar 2007
49. Agraz, F., Azodolmolky, S., Angelou, M., Perelló, J., Velasco, L., Spadaro, S., Francescon, A., Saradhi, C.V., Pointurier, Y., Kokkinos, P., Varvarigos, E., Gunkel, M., Tomkos, I.: Experimental demonstration of centralized and distributed impairment-aware control plane schemes for dynamic transparent optical networks. Optical Fiber Communications Conference (OFC 2010), PDPD5, San Diego, Mar 2010
50. Antona, J.-C., Bigo, S.: Physical design and performance estimation of heterogeneous optical transmission systems. C.R. Physique **9**(9–10), 963–984 (2008)
51. Chiu, A.L., Choudhury, G., Clapp, G., Doverspike, R., Gannett, J.W., Klincewicz, J.G., Li, G., Skoog, R.A., Stand, J., Von Lehmen, A., Xu, D.: Network design and architectures for highly dynamic next-generation IP-over-optical long distance networks. IEEE J. Lightw. Technol. **27**(12), 1878–1890 (2009)
52. Vereecken, W., Van Heddeghem, W., Puype, B., Colle, D., Pickavet, M., Demeester, P.: Optical networks: how much power do they consume and how can we optimize this. European Conference on Optical Communication (ECOC), Mo.1.D.1, Torino, Sept 2010
53. Kilper, D.C., Neilson, D., Stiliadis, D., Suvakovic, D., Goyal, S.: Fundamental limits on energy use in optical networks. European Conference on Optical Communication (ECOC), Tu.3.D.1, Torino, Sept 2010
54. Kilper, D.C., Atkinson, G., Korotky, S.K.: Optical transparency and network energy efficiency. IEEE International Conference on Transparent Optical Networks (ICTON 2010), Munich, June 2010
55. GreenTouch^{TM} consortium (www.greentouch.org)

Chapter 6
Conclusions and Outlook

In this book both efficient design of optical long-haul communication systems by novel algorithms as well as operation of optical core networks using new physical-layer impairment (PLI) aware routing algorithms have been presented. Furthermore, related aspects such as sparse optical–electrical–optical regenerator placement and optimization of core network energy efficiency have been examined.

For improving the network design phase two new methods have been developed: meta-heuristic based optimization and parallelization of a simulation on a multi core graphics processing unit. Each of these algorithms allows a speedup of more than 100 compared to conventional simulations used so far in optical communication systems modeling. Up to now grid search with equidistant sampling points to identify the optimum operating point of a system and CPU-based simulation utilizing only a single processing core has been state-of-the-art.

In the design phase of optical long-haul communication systems often a high number of parameters need to be optimized. Especially the power levels of the optical signals and the dispersion compensation map may be adapted on a span-by-span basis in a pre-defined parameter range with sometimes limited granularities (e.g. for the dispersion compensation modules). The meta-heuristic optimization method presented in this book has been shown to work well for such high dimensional optimization problems. To verify the accuracy of the proposed meta-heuristic optimization algorithm the results have been compared to several exemplary scenarios typical for optical long-haul communication systems. For these scenarios only a very low deviation of the results predicted by the meta-heuristic algorithm by a few percent from the theoretical optimum has been observed at a significantly reduced number of (time consuming) numerical simulations. In the investigated reference setups a typical reduction in the number of numerical simulations by a factor of more than 100 has been found.

Another promising way of reducing the computational time of numerical simulations is to parallelize them e.g. on a high number of processing cores.

In recent years graphics processing units (GPUs) have become a promising and relatively inexpensive platform for such applications. In this book a novel parallel implementation of the split-step Fourier method on a GPU has been presented. It can utilize the maximum available computational power of current GPUs, which is only available in single precision accuracy as most graphics cards are built for gaming purposes that do not require a high double precision performance. However, it has turned out that the accuracy of current GPUs especially of trigonometric functions in single precision is not high enough for some optical system simulations with a high number of fiber spans and a high number of WDM channels. This is why a novel implementation of the FFT in single precision based on precalculated twiddle factors with double precision accuracy has been presented in this book with an accuracy, which is even higher than the widely deployed FFTW (executed on a CPU) and at the same time exhibiting a high speedup compared to CPU-based simulations. This single precision based implementation of the split-step Fourier method shows a Q-factor penalty below 0.1 dB up to 100,000 split-steps, which is enough for most simulation scenarios.

Still, for some transmission system simulations requiring a very high number of split-steps, the use of single precision accuracy may not be sufficient. To offer a simulation scheme which can be utilized under these conditions, in this book a stratified Monte-Carlo sampling method has been proposed based on the combination of single and double precision calculations automatically determining whether a higher amount of double precision simulations is needed for a predefined accuracy. With this method a speedup of up to a factor of 180 has been observed compared to CPU-based standard Monte-Carlo simulations for a realistic transmission system setup with 20 spans and 112 Gb/s coherent polarization multiplexed transmission mixed with 10×10 Gb/s NRZ-OOK channels.

By using the two algorithms—meta-heuristic based optimization and GPU-based parallelization—presented in this book together, the individual speedups of the two methods can be joined, leading to an exceptionally high gain in computational time of more than 20,000.

Apart from numerical simulations of optical transmission systems also analytical and semi-analytical models for assessing the signal quality of a wavelength division multiplex optical transmission system with 10 Gb/s channels have been investigated. The use of such approximation models is mandatory, if the performance of the system needs to be assessed on-the-fly. This is needed e.g. during the operation of a transmission system when the optical system performance shall be included into the routing decision. The accuracy of the proposed models has been compared to laboratory experiments yielding a good agreement.

Most of the analytical models used for the assessment of the signal quality have been developed for 10 Gb/s NRZ-OOK channels. In this book possible extensions to higher bit rate channels with alternative modulation formats have been proposed. They are based on the combination of analytical models for linear degradation effects with numerical simulation of single channels to approximate the nonlinear intrachannel impairments. So far, however, no comprehensive set of analytical or semianalytical models for 112 Gb/s channels with advanced modulation formats is available. Furthermore, the consideration of electrical equalization algorithms, as are used for coherent polarization multiplexed transmission, will be very difficult to implement in such analytical models.

During the operation of optical core networks a dynamic and configurable optical layer, which is able to serve dynamic wavelength requests, is envisoned for the future [1]. Lightpaths (wavelengths with a large bandwidth capacity) may be requested by the user on demand. To setup such dynamic wavelength demands new PLI aware routing algorithms have been developed. In this book it has been shown that considering the expected signal quality in routing decisions can offer a significantly increased transparent reach (of e.g, more than a factor of 4 [2]) compared to shortest-path algorithms with fixed worst-case transparent distance constraints. This better approximation of the actual reach allows accommodating up to five times more traffic in a network compared to routing with worst-case estimations at the same blocking probability level. PLI-aware routing is especially beneficial in high network load scenarios. The reason is that long detours may be possible allowing to avoid highly utilized paths. These detours would not be feasible, if only a fixed maximum transparent distance criterion was used for the assessment of the minimum required signal quality.

The better utilization of the actual transparent reach can also be used to deploy a lower number of (expensive) optical-electrical-optical regenerators leading to a lower capital expenditure. It has been shown that the number of costly OEO regenerators can be reduced by 55% compared to fixed maximum reach based routing, if the signal quality along a path can be assessed during network operation on-the-fly. In this case only regenerators will be utilized, if the signal quality is not sufficient for the desired path. Additional improvements can be gained, if the current network utilization is included in the path assessment. Especially in partially loaded networks longer paths can be opened up due to the reduced amount of inter-channel crosstalk (compared to the case with full network load).

In the recent past the field of energy efficiency of network equipment has moved into the focus of many research efforts. Especially concerns have been raised that the energy consumption of information and communication technology equipment is lagging network traffic growth rates [3]. While the access networks currently consume the highest amount of energy, with rising traffic volume this is expected to shift to the core networks [4]. In this book the energy efficiency of core networks has been investigated. To improve energy efficiency an approach has been shown making use of load adaptive optical-bypassing and deactivation of (unused) network elements. Core networks need to be designed to meet peak traffic requirements. However, during the day the traffic load varies in a wide range with an absolute low in the early morning, where only roughly 25% network traffic is transported compared to the peak traffic. Therefore it is desirable to follow these traffic variations also with the energy consumption of the network components. In this book it has been investigated how the energy consumption of an IP over WDM optical core network can be decreased. Savings of more than 50% can be achieved by the deactivation of network components, if they become unused, in combination with the use of optical bypasses to keep more traffic in the optical layer.

In the future low price, high flexibility and robustness will become even more important than today. It will not only be crucial to reduce the cost of the network installation (CAPEX), but also the operational costs (OPEX) need to be minimized. This is especially important as the network traffic is growing exponentially with growth rates in the range of 50–100% p.a. However, it is not possible to merely scale the underlying transport architecture with the same speed as customers are not willing to accept a much higher price for the offered higher bandwidth. The network operators as well as the system vendors have to offer solutions which follow the high bandwidth growth rates without increasing the current price level too much. In this context especially increased flexibility of network operation is crucial, where constraint-based routing and dynamic setup of network resources on demand—as shown in this book—may be key elements.

An interesting feature may become the monitoring of the (optical) signal quality during network operation, too. Currently, the signal quality (i.e., the bit error ratio) is monitored only at the end of a path when this information is available from the forward error correction, and limited additional information from the optical layer such as the OSNR may be available. If more information about the signal degradation along a (transparent) transmission path is available, a better adaption to the current traffic situation will be possible. This leads to a better utilization of the network resources. The value of optical performance monitoring will rise with increasing transparency. The challenge will be to reduce the costs of optical performance monitors and to find the right amount of monitoring coverage [5].

Another focus of future research may be set on all-optical access solutions. Currently, the roll-out of a nationwide fiber optical access network is envisioned, offering much higher data rates than today's DSL solutions, for the next years [6]. There remain many issues to be solved. For example it is still an open question how optical access networks can be monitored efficiently (and preferably passively). Monitoring is desirable during all three parts of the lifecycle: during installation, for normal operation and in the case of fault localization and elimination. Currently, the most practical solution regarding technical and economical boundary conditions may be a combination of monitoring the alarms generated in the active network elements by the network management and of the implementation of passive monitoring (e.g. by OTDR) without traffic interruption [7]. Still it is a challenge to localize a failure in a PON with a high splitting ratio and with potentially very small differences in fiber lengths to the different users. Other activities focus on the expandability of such passive optical networks. PONs are currently investigated in WDM and TDM configurations both offering individual advantages. For the next generation of TDM PONs, bit rates of 2.5 or 10 Gb/s are proposed, which are however shared between all users. To meet the goal of providing a sustained data rate of about 1 Gb/s per user thus WDM PONs may be preferential [8]. Up to now it is unclear which solution will ultimately be deployed in the coming years.

The research in energy efficient networks has just begun recently. However, in this area significant efforts will be needed in the future. Currently, communication equipment uses only a one-digit figure of the total (global) energy consumption.

In the future, though, if no countermeasures are implemented, this will increase drastically due to the exponential growth of the traffic. If the growth rate is only 40% per year, which is a relatively conservative estimation, in 10 years an almost 30-times higher traffic volume needs to be processed. To improve the energy consumption significantly several steps have to be taken. At the end-user power saving measures have to be implemented to shut-down the line termination equipment (e.g. DSL modem or FTTH line terminal), if it becomes unused. In the metro and core area it will become crucial to keep as much traffic as possible in the (more energy efficient) optical layer without (electrically) processing data in the higher OSI layers. Dynamic reconfiguration of the network components following traffic patterns and deactivation of unused resources will significantly reduce the amount of energy consumed. There also exist radical proposals suggesting to use exclusively passive cooling for large computer and switching centers and favoring arctic regions as suitable locations. Moreover the energy efficiency on the component level has to be increased. To achieve these goals entirely new solutions may be needed leaving a vast field for future research activities.

References

1. Azodolmolky, S., Klinkowski, M., Marín Tordera, E., Careglio, D., Solé-Pareta, J., Tomkos, I.: A survey on physical layer impairments aware routing and wavelength assignment algorithms in optical networks. Elsevier J. Comput. Netw. **53**(7), 926–944 (2009)
2. Pachnicke, S., Luck, N., Krummrich, P.: Online physical-layer impairment-aware routing with quality of transmission constraints in translucent optical networks. In: IEEE International Conference on Transparent Optical Networks (ICTON 2009), Ponta Delgada, Portugal, June 2009
3. Kilper, D.C., Neilson, D., Stiliadis, D., Suvakovic, D., Goyal, S.: Fundamental limits on energy use in optical networks. In: European Conference on Optical Communication (ECOC), Tu.3.D.1, Torino, Sept 2010
4. Lange, C., Kosiankowski, D., Gerlach, C., Westphal, F.-J., Gladisch, A.: Energy consumption of telecommunication networks, European Conference on Optical Communications (ECOC), 5.5.3, Vienna, Sept 2009
5. Kilper, D.C., Bach, R., Blumenthal, D.J., Einstein, D., Landolsi, T., Ostar, L., Preiss, M., Willner, A.E.: Optical performance monitoring. IEEE J. Lightw. Technol. **22**(11), 294–304 (2004)
6. Teixeira, A.: Standardization in PONs: status and possible directions. In: IEEE International Conference on Transparent Optical Networks (ICTON 2010), Munich, June 2010
7. Erhard, A., Foisel, H.-M., Escher, F., Templin, A., Adamy, M.: Monitoring of the transparent fibre inrastructure for FTTx networks: An operator's view. In: IEEE International Conference on Transparent Optical Networks (ICTON 2010), Munich, June 2010
8. Rohde, H., Smolorz, S., Gottwald, E.: Next generation ultra high capacity PONs. In: IEEE Photonics Society Annual Meeting, WH1, Denver, Nov 2010

Appendix

Abbreviations

Acronym	Expansion
ADC	Analog-to-Digital Converter
ASE	Amplified Spontaneous Emission
ASIC	Application Specific Integrated Circuit
ASON	Automatically Switched Optical Network
ATD	Auto-Topology-Discovery
ATM	Asynchronous Transfer Mode
AWG	Arrayed Waveguide Grating
BER	Bit Error Ratio
BERT	Bit Error Ratio Tester
BP	Blocking Probability
BPP	Bin Packing Problem
CAPEX	Capital Expenditure
CBR	Constraint-Based Routing
CCI	Connection Controller Interface
CD	Chromatic Dispersion
CMA	Constant Modulus Algorithm
CP	Coherent Polarization Multiplex
CPU	Central Processing Unit
CRS	Carrier Routing System
CUDA	Compute Unified Device Architecture
CUFFT	CUDA^{TM} FFT
CW	Continuous Wave
CWDM	Coarse Wavelength Division Multiplexing
DAC	Digital-to-Analog Converter
DARPA	Defense Advanced Research Projects Agency
DB	Duobinary

(continued)

(continued)

Acronym	Expansion
DBR	Distributed Bragg Reflector
DCF	Dispersion Compensating Fiber
DCM	Dispersion Compensation Module
DDE	Decision Directed Equalizer
DEMUX	Demultiplexer
DFB	Distributed Feed Back
DFT	Discrete Fourier Transform
DGD	Differential Group Delay
DI	Delay Line Interferometer
DP	Double Precision
DPSK	Differential Phase Shift Keying
DQPSK	Differential Quadrature Phase Shift Keying
DSF	Dispersion Shifted Fiber
DSL	Digital Subscriber Line
DSP	Digital Signal Processing
DUCS	Distributed Under Compensation Scheme
DWDM	Dense Wavelength Division Multiplex
EDF	Erbium-Doped Fiber
EDFA	Erbium-Doped Fiber Amplifier
E-NNI	Exterior-Network Node Interface
EOP	Eye Opening Penalty
EPD	Electronic Pre-Distortion
ETDM	Electrical Time Division Multiplex
FBG	Fiber Bragg Grating
FCC	Fabric Card Chassis
FEC	Forward Error Correction
FFE	Feed Forward Equalizer
FIR	Finite Impulse Response
FFT	Fast Fourier Transform
FFTW	Fastest Fourier Transform in the West
FMAD	Fused Multiply and Add
FOM	Figure Of Merit
FOCS	Full-Inline and Optimized Post Compensation Scheme
FTTH	Fiber to the home
FWHM	Full Width at Half Maximum
FWM	Four-Wave Mixing
GaAs	Gallium Arsenide
G-MPLS	Generalized-Multi Protocol Label Switching
GPGPU	General Purpose computation on Graphics Processing Units
GPU	Graphics Processing Unit
GUI	Graphical User Interface
GVD	Group Velocity Dispersion
IETF	Internet Engineering Task Force
IF	Intermediate Frequency
IFFT	Inverse FFT

(continued)

Acronym	Expansion
I-FWM	Intra Channel Four-Wave Mixing
IIR	Infinite Impulse Response
InGaAsP	Indium Gallium Arsenide Phosphide
I-NNI	Interior-Network Node Interface
InP	Indium Phosphide
IP	Internet Protocol
IQ	Inphase and Quadrature
ISI	Inter Symbol Interference
ITU	International Telecommunication Union
IXP	Internet Exchange
I-XPM	Intra Channel Cross Phase Modulation
JP	Joint Polarization
KL	Karhunen-Loève
LCC	Line Card Chassis
LEAF	Large Effective Area Fiber
LHC	Left Hand Circular
LHD	Latin Hypercube Design
$LiNbO_3$	Lithium Niobate
LMS	Least Mean Squares
LO	Local Oscillator
MC	Monte-Carlo
MEMS	Micro Electro Mechanical Switches
MIMO	Multiple-Input-Multiple-Output
MLSE	Maximum Likelihood Sequence Estimation
MPLS	Multi Protocol Label Switching
MSC	Modular Services Card
MUX	Multiplexer
MZM	Mach-Zehnder Modulator
NLSE	Nonlinear Schrödinger Equation
NMI-A	Network Management Interface-A
NMI-T	Network Management Interface-T
NMS	Network Management System
NRZ	Non-Return to Zero
NVCC	NVIDIA® C Compiler
NZDSF	Non-Zero Dispersion Shifted Fiber
OCC	Optical Connection Controller
O-E-O	Optical-Electrical-Optical
OFDM	Orthogonal Frequency Division Multiplexing
ONU	Optical Network Unit
OOK	On-Off-Keying
OP	Outage Probability
OPEX	Operating Expenses
OPM	Optical Performance Monitoring
OSI	Open Systems Interconnection
OSNR	Optical Signal to Noise Ratio

(continued)

Acronym	Expansion
OTDM	Optical Time Domain Multiplexing
OTDR	Optical Time Domain Reflectometer
OXC	Optical Cross Connect
PBS	Polarization Beam Splitter
PDF	Probability Density Function
PIN	Positive Intrinsic Negative
PG	Pulse Generator
PLI	Physical Layer Impairments
PLIM	Physical Layer Interface Module
PMD	Polarization Mode Dispersion
POLMUX	Polarization Multiplex
PON	Passive Optical Networks
PRBS	Pseudo Random Bit Sequence
PSCRIPT	PHOTOSS Script
PSK	Phase Shift Keying
PSP	Principle State of Polarization
QAM	Quadrature Amplitude Modulation
QOS	Quality of Service
QPSK	Quadrature Phase Shift Keying
RHC	Right Hand Circular
RMSE	Root mean square error
RS	Reed Solomon
RWA	Routing and Wavelength Assignment
RX	Receiver
RZ	Return to Zero
SBS	Stimulated Brillouin Scattering
SC	Separated Channels
SDH	Synchronous Digital Hierarchy
SIMD	Single Instruction Multiple Data
SM	Streaming Multiprocessor
SMF	Single Mode Fiber
SNR	Signal to Noise Ratio
SONET	Synchronous Optical Network
SOP	State of Polarization
SOPMD	Second Order Polarization Mode Dispersion
SP	Single Precision
SPM	Self-Phase Modulation
SRS	Stimulated Raman Scattering
SS	Stratified Sampling
SSF	Split Step Fourier
SSFM	Split Step Fourier Method
SSMF	Standard Single Mode Fiber
STM	Synchronous Transport Module
TDM	Time Division Multiplexing

(continued)

Acronym	Expansion
TF	Total Field
TPC	Thread Processing Cluster
TWRS	Lucent True Wave RS (Reduced Slope)
TX	Transmitter
UNI	User Network Interface
VOA	Variable Optical Attenuator
VOIP	Voice-over-IP
WAN	Wide Area Network
WDM	Wavelength Division Multiplexing
XPM	Cross Phase Modulation

Fiber Parameters

Table A1 Parameters of the different fiber types at $\lambda = 1550$ nm

Fiber Type	Attenuation coefficient α (dB/km)	Dispersion constant D (ps/(nm·km))	Dispersion slope parameter S (ps/(nm^2·km))	Nonlin. index coeff. n_2 (m^2/W)	Eff.core A_{eff} (μm^2)	Nonlin. constant γ ((W·km)$^{-1}$)
SSMF						
Standard SMF	0.2	17	0.056	$2.7 \cdot 10^{-20}$	80	1.37
Alcatel ASMF	0.21	17	0.058	$2.7 \cdot 10^{-20}$	83	1.32
Lucent AllWave	0.23	16.6	0.088	$2.7 \cdot 10^{-20}$	80	1.26
OFS AllWave	0.21	17.3	0.058	$2.7 \cdot 10^{-20}$	86.6	1.26
Pure Silica	0.19	18	0.058	$2.7 \cdot 10^{-20}$	75	1.46
NZDSF+						
Alcatel TeraLight	0.23	8	0.058	$2.7 \cdot 10^{-20}$	65	1.68
Corning LEAF	0.23	4.2	0.06	$2.3 \cdot 10^{-20}$	72	1.29
Corning ELEAF	0.21	3.8	0.081	$2.3 \cdot 10^{-20}$	72	1.29
Lucent TrueWave RS	0.23	4	0.05	$2.7 \cdot 10^{-20}$	50	1.99
Lucent TrueWave+	0.23	3.7	0.065	$2.7 \cdot 10^{-20}$	55	1.99
OFS TW_RS	0.20	4.4	0.043	$2.7 \cdot 10^{-20}$	52.1	2.09
Pirelli Freelight	0.23	4.2	0.088	$2.7 \cdot 10^{-20}$	72	1.52
NZDSF−						
Lucent TrueWave−	0.23	−3	0.056	$2.7 \cdot 10^{-20}$	55	1.99
OFS SRS (submarine)	0.215	−3.1	0.05	$2.7 \cdot 10^{-20}$	50	2.18
OFS TW_XL	0.21	−3.0	0.112	$2.7 \cdot 10^{-20}$	72.0	1.52
Corning MetroCore	0.25	−7.6	0.12	$2.7 \cdot 10^{-20}$	51.5	2.12
Corning LS	0.25	−1.3	0.071	$2.7 \cdot 10^{-20}$	55	1.99
DSF						
DSF	0.23	0	0.056	$2.7 \cdot 10^{-20}$	55	1.99
DCF						
Lycom DK	0.5	−102	−0.23	$2.2 \cdot 10^{-20}$	17	5.24
Lycom WBDK	0.5	−102	−0.35	$2.2 \cdot 10^{-20}$	17	5.24
Lycom HSDK	0.5	−102	−1.1	$2.2 \cdot 10^{-20}$	17	5.24
Corning SC	0.5	−111	−2.335	$2.2 \cdot 10^{-20}$	17	5.24

Index

Symbols
16-QAM, 17
90°-hybrid, 13, 17, 25

A
Absorption
  intrinsic, 32
Amplified spontaneous emission (ASE) *noise*
Analog-to-digital converter (ADC), 17, 27
Analytical model, 57, 84
Application specific integrated circuit
  (ASIC), 26
Arrayed waveguide grating (AWG), 12, 14, 24
Artificial four-wave mixing, 45, 75, 77
Attenuation, 18, 32, 41, 44, 88, 127
Automatically switched optical network
  (ASON), 95, 106, 108, 120
Auto-topology-discovery, 109

B
Backward error, 74
Balanced receiver, 17, 50
Baseband, 24
  complex, 31
Beat length, 41
Bias current, 13
Bin packing problem (BPP), 132
Birefringence *fiber birefringence*
Bit error ratio (BER), 26, 37, 48–49,
  51, 82, 87
Bit rate, 11, 17, 19, 34, 57, 83, 125, 132
Blocking probability (BP), 118–120, 123,
  128, 134

C
Capacity-distance product, 3
Capital expenditure (CAPEX), 117, 130, 149
Central processing unit (CPU), 4, 55, 68, 72,
  81, 97
Cladding, 18
Coherent detection, 3, 13, 17, 20, 50
Complex envelope, 31
Compute Unified Device Architecture
  (CUDA), 68, 74
Conduction band, 24
Connection request, 106, 110, 116, 120
Constant modulus algorithm (CMA), 27
Constellation diagram, 15, 27
Constraint-based routing (CBR), 109, 111,
  119, 127
Continuous-wave (CW), 13, 90–91
Control plane, 105, 108, 142
Cooley–Tukey algorithm, 69
Core router, 131
Coupled-mode equation, 41
Cross section, 32, 36
  absorption, 36
  emission, 36
Crosstalk, 26, 39–41, 44, 84, 91, 112

D
DCF granularity, 67, 119, 128
Decision directed equalizers (DDE), 26
Decision threshold, 49
Delay line interferometer (DI), 16, 24
Demand granularity, 132
Demand scaling factor, 110, 120, 123, 128
Demultiplexing, 24–26

D (cont.)
Detection, 16
  direct, 16, 24, 49
  heterodyne, 24
  homodyne, 24–25
  single-ended, 25, 50
Differential group delay (DGD), 85, 112, 126
Differential phase shift keying (DPSK), 12, 15
Differential quadrature phase shift keying (DQPSK)
  quadrature phase shift keying (QPSK), 12, 14, 17, 81, 125
Digital signal processing (DSP), 26, 50
Digital subscriber line (DSL), 130
Digital-to-analog converter (DAC), 17
Discrete Fourier transform (DFT), 69
  1D DFT, 71
Dispersion, 18, 32
  dispersion compensation, 19, 34
  dispersion map, 63, 67
  dispersion slope, 19, 34
  distributed under-compensation scheme (DUCS), 20
  fourth order, 34
  full-inline optimized post-compensation scheme (FOCS), 20, 92
  group velocity dispersion (GVD), 33, 39, 84, 111
  higher order PMD, 86
  polarization mode dispersion (PMD), 38, 46, 84, 111, 120, 122
  pre-compensation, 20
  second order, 34
  third order, 34
Dispersion compensating fiber (DCF)
  dispersion
Double precision (DP), 68, 72, 76, 81
Duobinary, 15, 64, 125–126

E
Edge weight, 114
Electrical time division multiplexing (ETDM), 14
Energy consumption, 4, 105, 129, 136
Energy efficiency, 4, 106, 129
Erbium-doped fiber amplifier (EDFA), 3, 18, 21, 35, 48, 87, 112, 126
Extinction ratio, 87
Eye opening, 51, 87
  penalty, 64, 84

F
Fabric card chassis (FCC), 131
Fast Fourier transform (FFT), 44, 68
  auto-tuning algorithm, 71
  CUFFT, 68, 74–75
  FFTW library, 74
  four-step FFT, 74
  IPP library, 74
  optimum factorization, 72
  out-of-place algorithm, 71
  radix, 71
Feed-forward equalizers (FFE), 26
Fiber birefringence, 38, 41, 46–47
  large birefringence, 41
  low birefringence, 38
Fiber Bragg grating (FBG), 12, 34
Fiber parameters, 157
Fiber-to-the-home (FTTH), 130
Filter, 13, 26, 49, 64, 89, 111
  gain flattening, 18, 21, 94
Filter crosstalk, 112
  intra-band, 113
  out-of-band, 113
Finite impulse response (FIR), 26
Forward error correction (FEC), 26, 48, 116
Frequency domain, 44, 47, 75, 88
Fundamental mode, 38
Fused multiply and add (FMAD), 72

G
Generalized Multi-Protocol Label Switching (GMPLS), 106
Graph, 117, 136
Graphics processing unit (GPU), 6, 68, 71, 75
Grid computing, 57, 97
Grid search, 63, 66
Grooming, 132

H
Heterostructure, 24
  double, 24
Heuristic, 56, 116, 119, 125, 135
Higher-order modulation, 17
Holding time, 110
Hop, 133

I
IEEE-754, 69, 72
Infinite impulse response (IIR), 26, 44
Interarrival time, 109
Intermediate frequency (IF), 25

Index

Internet Engineering Task Force (IETF), 106
Internet exchange, 1
 DE-CIX node, 1
Internet protocol (IP), 106, 108, 110, 130
Interpolation function, 56, 59
 equalization plane, 60
 Hardy multiquadric, 59
 radial basis, 60
 scattered data, 59
Intrachannel effects *nonlinear fiber effects*
IP-over-WDM, 130
IQ modulator, 14
ITU-T, 26, 108

J
Jones vector, 46

K
Karhunen-Loève (KL), 50
Kerr effect, 39
k-shortest paths, 115

L
Laser, 11, 13, 21, 40, 76
 distributed Bragg reflector (DBR), 13
 distributed feedback (DFB), 13
Latin hypercube design (LHD), 58
 maximin design , 59
Least-mean squares (LMS), 27
Line card chassis (LCC), 131
Liquid crystal, 22
Load-adaptive operation, 138
Local oscillator (LO), 24, 27
Lookup table, 74, 115

M
Mach–Zehnder modulator (MZM), 11, 14
Maximum likelihood sequence estimation (MLSE), 26
Maximum-likelihood estimator, 79
Meta-model, 55
 pseudo-code, 62
Micro-electro-mechanical system (MEMS), 22
Modulation format, 11, 14, 57, 83, 120
Monte Carlo (MC), 37, 52, 83
 confidence level, 79
 fast system, 81
 stratified sampling, 78
 true system, 81

Multiple-input-multiple-output (MIMO), 27
Multiplexing, 11, 18

N
Network management, 108
 network management system (NMS), 112
Noise, 21
 amplified spontaneous emission (ASE), 21, 27, 35, 87, 94, 111, 127
 ASE-ASE beat noise, 36, 49
 ASE-channel noise, 49
 ASE-shot noise, 36, 49
 bins, 38
 noise figure, 21, 36, 66, 75, 87, 112, 120, 122, 126
 power density, 37, 49
 shot-noise, 35–36, 48
Nonlinear coefficient, 41
Nonlinear fiber effects, 39, 94
 cross-phase modulation (XPM), 39, 88, 111, 119, 125
 four-wave mixing (FWM), 40, 91, 111, 125
 intrachannel cross-phase modulation (I-XPM), 41
 intrachannel four-wave mixing (I-FWM), 40
 polarization crosstalk (XPolM), 40
 Raman effect, 21
 self-phase modulation (SPM), 39, 113
 stimulated Brillouin scattering (SBS), 40
 stimulated Raman scattering (SRS), 40, 94
Nonlinear phase shift, 45
Nonlinear Schrödinger equation (NLSE), 32, 42, 75, 83
 coupled NLSE, 43
Non-return to zero (NRZ), 11, 14, 57, 63, 75, 84, 88, 92, 119, 125

O
Offline routing, 113–114, 120
OH^- absorption, 33
Online routing, 113, 115, 124
On-off keying (OOK), 11, 15, 57, 81, 90, 111, 119, 125
Open Systems Interconnection (OSI), 130
Operational expenditure (OPEX), 4, 117, 129
Operator splitting, 45
 asymmetrical, 45
 symmetrical, 45
Optical bypassing, 133–134
Optical cross connect (OXC), 11, 22, 105, 113, 132, 136

O (*cont.*)
Optical phonon, 40
Optical signal-to-noise ratio (OSNR), 35, 51, 87, 112, 126
Optical time domain multiplexing (OTDM), 14
Optical transmission system *transmission system*
Optical-electrical-optical (OEO), 23, 105, 114, 121, 130
Optimization, 55, 97
Orthogonal frequency division multiplexing (OFDM), 17
Outage probability, 87

P
Parallelization, 55
Parameter space, 56
Phase locked loop, 25
Phase matching, 92
Photodiode, 16, 24, 36, 49
  PIN, 24, 48
PHOTOSS, 31, 45, 50, 97
  PScript, 52
Physical layer, 51, 105–106, 114, 121, 129
Physical-layer impairment based routing *constraint-based routing*
pn-junction, 13, 24
Poisson process, 109
Polarization beam splitter, 14, 22, 25
Polarization crosstalk, 39
Polarization mode dispersion (PMD) *dispersion*
Polarization multiplexing (polmux), 12, 14, 24, 40, 83
Population model, 118, 135
Power consumption, 4, 106, 136
Power spectral density, 89
Principle state of polarization (PSP), 86
Probability density function, 50, 109
Propagation constant, 32, 38
Pseudo-random bit sequence (PRBS), 63, 76, 89, 92
Pump-probe model, 88

Q
Q-factor, 51, 76, 84, 87, 94, 111–112, 114, 121
Q-penalty, 113, 127
Quadrature amplitude modulation (QAM), 17, 57

Quadrature phase shift keying (QPSK), 12, 14, 17, 57, 82, 125
Quantum efficiency, 35

R
Radial basis function, 60
Raman effect
  gain factor, 43
  Raman pumping, 12, 21
  time constants, 42
Rayleigh scattering, 32
Recirculating loop, 90, 92, 111
Reduction of power consumption, 137
Reference topology, 110, 118, 121, 125, 134
Refractive index, 13, 18, 32, 39, 88
Regeneration, 23, 107, 116–117, 128
Regenerator placement, 116, 118, 120
Responsivity, 49, 113
Return to zero (RZ), 11–12, 84
Routing and wavelength assignment (RWA), 109, 114

S
Scattering, 40
  elastic, 40
  inelastic, 40
Separated channels (SC) model, 42
Shortest path algorithm, 114, 127
Signal-to-noise ratio (SNR), 36
  optical *optical signal-to-noise ratio (OSNR)*
Simulated annealing, 57
Single instruction multiple data (SIMD), 68, 75
Single mode fiber, 12, 18, 33, 46
  dispersion compensating fiber (DCF), 19
  dispersion shifted fiber (DSF), 19, 40
  non-zero dispersion shifted fiber (NZDSF), 19
  standard single mode fiber (SSMF), 19, 41, 64, 75, 88, 119
  step-index, 18
Single precision (SP), 68, 74–75
Soliton, 39
  dispersion managed soliton, 41
Spectral efficiency, 3, 12, 17
Spectral width, 34, 39, 49
Speedup factor, 77, 78
Split-step Fourier method (SSFM), 4, 32, 44, 56, 67, 75, 97, 127
Spontaneous emission, 21, 33, 35–36
Step-size, 75, 83, 127

alternating, 45
  maximum, 45
Stokes vector, 46
Stokes wave, 40
Streaming multiprocessor (SM), 69

T
Tail extrapolation, 48
Taylor series, 33, 72
Three-tone product, 92
Time domain, 44, 75, 84
Total field (TF) model, 42
Total probability law, 80
Traffic, 109
  traffic load, 110, 135, 138
  traffic matrix, 118
  traffic mix, 110
  traffic model, 109
Transfer function, 24, 88
Transient effects, 94, 139
Translucent, 106, 120
Transmission system, 11, 16, 20, 31, 48, 63, 67, 79, 84, 97, 105, 125
Transmission window
  C-band, 18, 34, 36
  L-band, 18
  S-band, 18
  U-band, 18
Transparency, 109, 133

Transparent reach, 11, 118–119, 121, 126, 127, 136
Twiddle factor, 71
Two-tone product, 92

U
User network interface (UNI), 109
Utilization, 74, 123, 131, 133, 135

V
Valence band, 24
Variable optical attenuator (VOA), 22
Viterbi-and-Viterbi algorithm, 27
Voice-over-IP (VoIP), 2, 130

W
Warm-up period, 111
Wave plate model, 47
Wavelength continuity, 114, 120
Wavelength division multiplex (WDM), 2, 11, 18, 39, 42, 56, 63, 67, 75, 88
Wavelet collocation method, 44

X
Xtalk *crosstalk*

Printed by Printforce, the Netherlands